W0235396

Agroforestry as Climate Change Adaptation

Mette Fog Olwig ·
Aske Skovmand Bosselmann · Kwadwo Owusu
Editors

Agroforestry as Climate Change Adaptation

The Case of Cocoa Farming in Ghana

Editors
Mette Fog Olwig
Department of Social Sciences
and Business
Roskilde University
Roskilde, Denmark

Aske Skovmand Bosselmann
Department of Food and Resource
Economics
University of Copenhagen
Frederiksberg, Denmark

Kwadwo Owusu
Department of Geography
and Resource Development
University of Ghana
Legon, Ghana

ISBN 978-3-031-45634-3 ISBN 978-3-031-45635-0 (eBook)
https://doi.org/10.1007/978-3-031-45635-0

© The Editor(s) (if applicable) and The Author(s) 2024. This book is an open access
publication.

Open Access This book is licensed under the terms of the Creative Commons Attribution
4.0 International License (http://creativecommons.org/licenses/by/4.0/), which permits
use, sharing, adaptation, distribution and reproduction in any medium or format, as long
as you give appropriate credit to the original author(s) and the source, provide a link to
the Creative Commons license and indicate if changes were made.
The images or other third party material in this book are included in the book's Creative
Commons license, unless indicated otherwise in a credit line to the material. If material
is not included in the book's Creative Commons license and your intended use is not
permitted by statutory regulation or exceeds the permitted use, you will need to obtain
permission directly from the copyright holder.
The use of general descriptive names, registered names, trademarks, service marks, etc.
in this publication does not imply, even in the absence of a specific statement, that such
names are exempt from the relevant protective laws and regulations and therefore free for
general use.
The publisher, the authors, and the editors are safe to assume that the advice and information in this book are believed to be true and accurate at the date of publication. Neither
the publisher nor the authors or the editors give a warranty, expressed or implied, with
respect to the material contained herein or for any errors or omissions that may have been
made. The publisher remains neutral with regard to jurisdictional claims in published maps
and institutional affiliations.

Cover illustration: © Helmut Schwigon

This Palgrave Macmillan imprint is published by the registered company Springer Nature
Switzerland AG
The registered company address is: Gewerbestrasse 11, 6330 Cham, Switzerland

Paper in this product is recyclable.

Acknowledgements

Special thanks to the Ghanaian farmers, village leaders and officials for agreeing to participate in our data collection and research, for always being helpful and welcoming, and for their invaluable insights. Additionally, we would like to thank our colleagues at the University of Copenhagen, in particular Søren Marcus Pedersen for good and useful feedback on an earlier draft, as well as Christian Pilegaard and Christian Lund for their astute comments to an earlier presentation of our work. We also thank our editor Rachael Ballard for her valuable inputs, support and guidance. Finally, we are grateful for the financial support received from Danida without which this study would not have been possible. The work was supported by Danida [Ministry of Foreign Affairs of Denmark] through "Climate Smart Cocoa Systems for Ghana, CLIMCOCOA", DFC project no: 16-P02-GHA.

CONTENTS

CONTRIBUTORS

Christiana A. Amoatey Department of Crop Science, University of Ghana, Accra, Ghana

Richard Asare International Institute of Tropical Agriculture (IITA), Accra, Ghana

Bismark Kwesi Asitoakor Department of Crop Science, University of Ghana, Accra, Ghana;
Department of Geosciences and Natural Resource Management, University of Copenhagen, Frederiksberg, Denmark;
CSIR-Plant Genetic Resources Research Institute, Bunso, Ghana

Sylvester Afram Boadi Department of Food and Resource Economics, University of Copenhagen, Frederiksberg, Denmark;
Department of Geography and Resource Development, University of Ghana, Accra, Ghana;
CSIR-Water Research Institute, Accra, Ghana

Aske Skovmand Bosselmann Department of Food and Resource Economics, University of Copenhagen, Frederiksberg, Denmark

Henrik Meilby Department of Food and Resource Economics, University of Copenhagen, Frederiksberg, Denmark

Eric Opoku Mensah Department of Crop Science, University of Ghana, Accra, Ghana;

Department of Geosciences and Natural Resource Management, University of Copenhagen, Frederiksberg, Denmark;
CSIR-Plant Genetic Resources Research Institute, Bunso, Ghana

Mette Fog Olwig Department of Social Sciences and Business, Roskilde University, Roskilde, Denmark

Kwadwo Owusu Department of Geography and Resource Development, University of Ghana, Accra, Ghana

Hans Peter Ravn Department of Geosciences and Natural Resource Management, University of Copenhagen, Frederiksberg, Denmark

Anders Ræbild Department of Geosciences and Natural Resource Management, University of Copenhagen, Frederiksberg, Denmark

Philippe Vaast UMR Eco & Sols, Centre de Coopération Internationale en Recherche Agronomique Pour Le Développement (CIRAD), Université Montpellier, Montpellier, France;
World Agroforestry Center (ICRAF), Nairobi, Kenya

ABBREVIATIONS

CHED	Cocoa Health and Extension Division
CIRAD	Centre of International Cooperation on Agricultural Research for Development
CLIMCOCOA	Climate Smart Cocoa System for Ghana
COCOBOD	Ghana Cocoa Board
CRU TS	Climatic Research Unit Timeseries
CRU	Climatic Research Unit
CSA	Climate-Smart Agriculture
CSC	Climate-Smart Cocoa
CSCWG	Climate Smart Cocoa Working Group
CSDDD	Corporate Sustainability Due Diligence Directive
CSSV	Cocoa Swollen Shoot Virus
Danida	Denmark's development cooperation under the Danish Ministry of Foreign Affairs
ERPAs	Emission Reductions Payment Agreements
EU	European Union
FAO	Food and Agriculture Organization
FAOSTAT	Food and Agriculture Statistics
FOB	Free On Board
GCFRP	Ghana Cocoa Forest REDD+ Program
GDP	Gross Domestic Product
GHS	Ghana Cedi
ICCO	International Cocoa Organization
ICRAF	World Agroforestry Centre
IITA	International Institute of Tropical Agriculture
IPCC	Intergovernmental Panel on Climate Change

IPM	Integrated Pest Management
LID	Living Income Differential
MoFA	Ministry of Food and Agriculture
NCCAS	Ghana National Climate Change Adaptation Strategy
NGOs	Non-Governmental Organizations
REDD	Reducing Emissions from Deforestation and Forest Degradation
SWC	Soil water content
UN	United Nations
UNFCCC	United Nations Framework Convention on Climate Change
USD	United States Dollar
WCF	World Cocoa Foundation

LIST OF FIGURES

LIST OF TABLES

CHAPTER 1

Introduction: Climate, Cocoa and Trees

*Mette Fog Olwig◉, Richard Asare◉, Henrik Meilby◉,
Philippe Vaast◉, and Kwadwo Owusu◉*

Abstract Climate change is predicted to significantly reduce areas suitable for the cultivation of cocoa, an important cash crop providing a livelihood to over six million smallholders in the humid tropics. Cocoa agroforestry shows potential to increase climate resilience while providing more stable incomes, enhancing biodiversity, supporting healthy ecosystems and reducing the pace at which farms expand into forested areas. Based on the multidisciplinary 'Climate Smart Cocoa Systems for Ghana' research project, this book investigates the case of the biophysical and socioeconomic sustainability of cocoa agroforestry in Ghana, the second largest producer of cocoa in the world. After a brief introduction to the research project, this introductory chapter reviews the literature on the links between climate change, farming and agroforestry, thereby situating

M. F. Olwig (✉)
Department of Social Sciences and Business, Roskilde University, Roskilde, Denmark
e-mail: mettefo@ruc.dk

R. Asare
International Institute of Tropical Agriculture (IITA), Accra, Ghana
e-mail: r.asare@cgiar.org

© The Author(s) 2024
M. F. Olwig et al. (eds.), *Agroforestry as Climate Change Adaptation*,
https://doi.org/10.1007/978-3-031-45635-0_1

1

the study within a wider context. It then presents an in-depth analysis of historical Ghanaian cocoa yields and climate data at both the national and regional levels to establish a foundation for understanding the new climate risks faced by cocoa farmers. The chapter concludes by providing an overview of the chapters that follow and introducing the overall argument that agroforestry can only successfully address climate change impacts on cocoa farming if location-specific biophysical and socioeconomic factors are considered.

Keywords Cocoa systems · Agroforestry · Climate-smart agriculture · Sustainable cocoa · Historical yield and climate data · Smallholders

1.1   Introduction

Cocoa is not only the key ingredient in chocolate, it is also an important cash crop providing a livelihood to over six million smallholder farmers in the humid tropics. It is cultivated on an estimated area of about 11.54 million ha in over sixty countries (FAOSTAT, 2021). However, being particularly sensitive to drought and high temperatures, the area suitable

H. Meilby
Department of Food and Resource Economics, University of Copenhagen, Frederiksberg, Denmark
e-mail: heme@ifro.ku.dk

P. Vaast
UMR Eco & Sols, Centre de Coopération Internationale en Recherche Agronomique Pour Le Développement (CIRAD), Université Montpellier, Montpellier, France

World Agroforestry Center (ICRAF), Nairobi, Kenya

P. Vaast
e-mail: philippe.vaast@cirad.fr

K. Owusu
Department of Geography and Resource Development, University of Ghana, Accra, Ghana
e-mail: kowusu@ug.edu.gh

for cocoa cultivation is predicted to decline substantially in the coming years due to climate change, with serious consequences for farmers' livelihoods and the cocoa industry. The potential of growing certain crops under shade, in particular coffee, as a form of agroforestry, has been given much attention as it is believed to be more climate-resilient, and hence more sustainable than growing these crops in the open (Vaast et al., 2016). This book investigates both the biophysical and socioeconomic sustainability of agroforestry in relation to cocoa in times of climate change. It focuses on Ghana, the second largest producer of cocoa in the world.

Cocoa agroforestry entails the growing of cocoa together with shade trees and food crops for agronomic, economic and environmental benefits. Cocoa agroforestry is thus part of a larger trend to encourage forestry as a tool for climate change mitigation and adaptation. Thus far, research on cocoa agroforestry is generally positive regarding its potential to increase farms' resilience to climate change while providing additional, diversified and more stable incomes, enhancing biodiversity, supporting healthy ecosystems and reducing the pace at which farms expand into forested areas (Andres et al., 2018; Asare et al., 2014, 2019; Blaser et al., 2018; Djokoto et al., 2017). Yet, in practice, it is difficult to implement cocoa agroforestry, because integrating shade trees into cocoa farming systems is no simple matter. It requires a good understanding of institutional and social factors, such as land- and tree-use rights and differentiated access to inputs and training. It is also crucial to have locally specific biophysical and socioeconomic knowledge of crop combinations and the types and densities of tree species that, when configured properly, improve complementarity and minimize competition for resources (nutrients, water, solar radiation), manage pests and diseases efficiently, and enhance yields of cocoa, as well as timber, firewood, fruits and other non-timber products. Cocoa agroforestry research, however, often focuses mainly on the health of cocoa, the export crop, and pays little attention to the complex interaction of different plant species and their environmental and societal attributes (Vaast & Somarriba, 2014).

This book addresses these gaps to better inform research, policy and practice. It does so by providing a comprehensive and novel understanding of agroforestry and cocoa production under changing climates through:

1. analysis of historical data on cocoa yields and climate in Ghana

2. on-farm studies and controlled experiments investigating the impact of not only shade levels, but different shade tree species on key factors such as pests and diseases, cocoa ecophysiology and cocoa yields

3. analysis of quantitative and qualitative data that elucidate the socio-economic factors influencing cocoa farmers' ability and willingness to adopt cocoa agroforestry, paying attention to the importance of particular shade tree species, as well as shade tree species diversity.

The book thus provides a multidisciplinary perspective on the potential of trees to mitigate the negative impacts of climate change through agroforestry.

Focusing on cocoa agroforestry in Ghana, the book compares findings across a climate gradient from the wet southern to the dry northern parts of the Ghanaian cocoa belt (see Fig. 1.1). The book has three key aims. First, it shows how agroforestry can provide a viable and profitable pathway for addressing the impacts of climate change. Second, it demonstrates the need to pay careful attention to context-specific socioeconomic and biophysical factors to maximize the potential of agroforestry and avoid unintended social and environmental consequences. Third, it demonstrates why multidisciplinary approaches are essential when studying climate change and agricultural sustainability.

This introductory chapter begins by providing a brief introduction to cocoa in Ghana and the multidisciplinary research project 'Climate Smart Cocoa Systems for Ghana (CLIMCOCOA)' from which this book emerges. This is followed by a review of the literature on the links between climate change, farming and agroforestry to situate the study in the broader literature on agroforestry and climate change. It then presents an in-depth analysis of historical cocoa yields and climate data both nationally and regionally to establish a foundation for understanding the new climate risks cocoa farmers must overcome to ensure future sustainable cocoa production in Ghana. The chapter concludes by providing an overview of the chapters that follow.

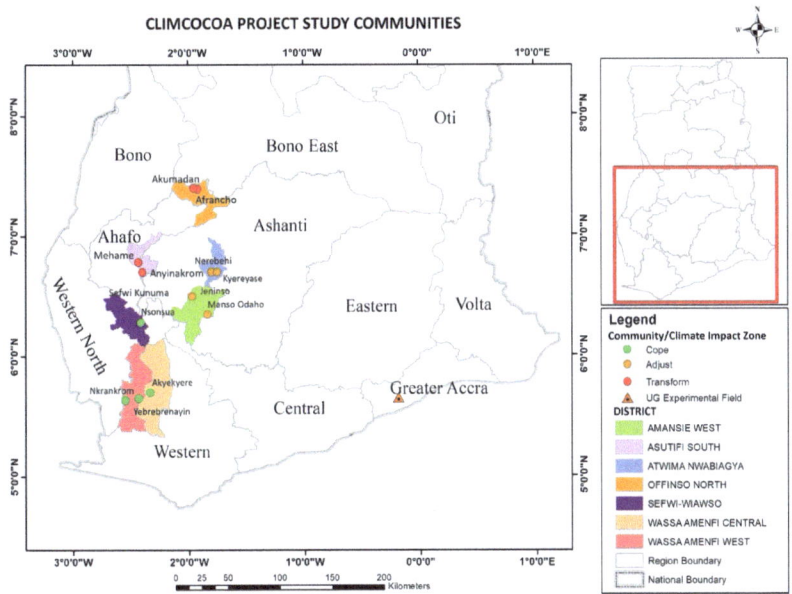

Fig. 1.1 Map of southern Ghana showing the CLIMCOCOA project's study communities

1.2 COCOA IN GHANA AND THE MULTIDISCIPLINARY RESEARCH PROJECT CLIMCOCOA

This book is based on research conducted between 2016 and 2021 as part of a comprehensive research project entitled 'Climate Smart Cocoa System for Ghana (CLIMCOCOA),' funded by the Danish Ministry of Foreign Affairs/Danida. The project team comprised researchers from the International Institute of Tropical Agriculture (IITA), Ghana; the World Agroforestry Centre (ICRAF), Kenya; the Centre of International Cooperation on Agricultural Research for Development (CIRAD), France; the University of Ghana; the University of Copenhagen, Denmark; and Roskilde University, Denmark. These researchers covered multiple disciplines, including human geography, climatology, development studies, natural resource economics, socioeconomics, ecophysiology, agroforestry, biometry and entomology.

The CLIMCOCOA research project focused on the case of Ghana. Ghana is important to the global production of cocoa, being the second largest producer in the world, and cocoa is also of key importance to Ghanaian society. Cocoa is one of Ghana's main exports making up 3.9% of Ghana's GDP in 2019 according to FAO statistics, while Sadhu et al. (2020) estimate the contribution much higher, at 7% of GDP. The cocoa sector furthermore employs 17% of the labor force, supporting the livelihoods of more than 550,000 farming households (Sadhu et al., 2020). Cocoa thus represents an important pillar in both rural and urban poverty alleviation and the general development of the Ghanaian economy.

Cocoa was introduced to Ghana in the nineteenth century and was traditionally established as a form of agroforestry on partially cleared forestland by smallholders. Cocoa arrived together with British colonial rule, which was keen to support the development of tradeable commodities. Since cocoa is generally not consumed locally, cocoa farming led to a reorientation from subsistence to labor-intensive cash-cropping, thereby exposing farmers to the risks of external market forces. This, coupled with the increasing use of seasonal migrant workers from the northern, poorer and drier part of Ghana and the use of hazardous and illicit child labor, meant that cocoa farming led to major changes in social, generational and gender relations (Allman, 1994; Sadhu et al., 2020; Yaro et al., 2021). In the 1980s, there was a dramatic drop in cocoa output attributed in part to the El Niño weather phenomenon, which led to a period of severe drought, and bushfires, that destroyed cocoa farms (Kolavalli & Vigneri, 2011). As a response, full-sun cocoa systems were introduced by the government coupled with new and early maturing cocoa varieties that thrived under less shade and in the short term produced higher yields than shaded cocoa (Gockowski et al., 2013). This resulted in the widespread adoption of full-sun cocoa systems by smallholders. Today, for a host of reasons that will be explored further in this book, researchers, extension officers and policymakers are increasingly favoring agroforestry instead of full-sun cocoa systems. This is in part because of financial considerations, as high yields from full-sun systems depend on a high level of inputs, which are expensive and can be difficult to obtain. Environmental concerns are also important because full-sun systems are established at the expense of forestlands, soil fertility, biodiversity and environmental sustainability. The current interest in Ghana, and in West Africa more generally, in reintroducing agroforestry systems is also a result of the new risks posed by climate change. Research indicates that agroforestry may be more resilient than full-sun systems when the correct level of shade and the appropriate combination of shade tree species are implemented. For example, this growing interest was apparent at the 2022 International Symposium on Cocoa Research in Montpellier, France (https://www.isc rsymposium.org/oral-presentations).

The CLIMCOCOA research project had two objectives: (1) to develop a comprehensive understanding of the impacts of climate change on the socio-biophysical bases of cocoa systems in Ghana; and (2) to assess the role of agroforestry as a model for climate-smart cocoa production. It utilized a multidisciplinary approach to investigate the biophysical and socioeconomic opportunities for and limitations of cocoa agroforestry under climate change. The project therefore employed multiple and varied methods, including analysis of historical yields and climate data; on-farm studies and eco-physiological experiments; literature reviews; field observations; twenty focus-group discussions and a household survey of 402 households in twelve cocoa communities. Data were collected in the Ashanti, Ahafo, Western and Western North Regions across the three delineated climate impact zones projected to have different degrees of climate suitability for cocoa production in Ghana (see Fig. 1.1). These three 'climate impact zones' are categorized as (1) the Cope Zone, which has the most favorable climate currently for cocoa and the lowest climate-related vulnerability, indicating the ability for cocoa farming to *cope* with climate change; (2) the Adjust Zone, with a moderately favorable current climate and moderate climate vulnerability, indicating the need for cocoa farming to make some *adjustments* to cope with climate change; and (3) the Transform Zone, which currently has the least favorable climate and the greatest climate vulnerability, indicating the need to replace or radically *transform* cocoa farming (Bunn et al., 2019).

This book presents the key findings from the CLIMCOCOA research project and discusses their implications for the future of cocoa cultivation in this current era of climate change.

1.3   CLIMATE CHANGE, FARMING AND AGROFORESTRY

The United Nations Framework Convention on Climate Change (UNFCCC) attributes climate change directly or indirectly to anthropogenic activities that change the composition of the global atmosphere leading to climatic changes that exceed the natural climate variability observed over comparable time periods (https://unfccc.int/). Similarly, the Intergovernmental Panel on Climate Change (IPCC) defines climate change as a change in the state of the climate identifiable by alterations in the mean and/or the variability of its properties persisting for an extended period—typically decades or longer (IPCC, 2013). Moreover, climate variability is attributed to all temporal and spatial scales beyond that of individual weather events in terms of the deviations of climatic statistics over a given period (e.g., a month, season or year) from the long-term statistics relating to the corresponding calendar period. In effect, climate variability is measured by deviations, which are usually termed anomalies.

Climate change adversely affects the conditions under which agricultural production in sub-Saharan Africa operates. In this region, as in other areas around the world, plants, animals and ecosystems are all experiencing the impact of the ongoing changes in climatic conditions. Some of these impacts, such as the direct impact of heat waves, droughts and floods, are affecting value chains of specific commodities in specific stages of the cultivation cycle. The effects of some of these impacts can be predicted with high confidence, whereas other impacts, such as the effect of climatic change on a whole ecosystem, are more complex to predict, since each element may react differently and interact with the others.

Cocoa is highly sensitive to changes in climatic conditions like drought and high temperatures (Ameyaw et al., 2018; Schroth et al., 2016). Changes in climatic conditions such as rainfall distribution and temperature fluctuations affect evapotranspiration and abiotic stress. According to Anim-Kwapong and Frimpong (2008), climate change can alter the development and incidence of cocoa diseases and pests, modify host tolerance and result in changes in the host's interactions with pests and diseases. This, the authors argue, could shift the geographical distribution of host–pathogen/pest interactions with negative effects on yields, subsequently affecting socioeconomic variables such as farm incomes, livelihoods and farm-level decision-making. Climate change, combined with the use of poor planting material, low soil-fertility management, the prevalence of diseases and pests, and the limited adoption of good agricultural practices have led to average cocoa yields stagnating in the four largest cocoa-producing countries in West Africa (Cameroon, Côte d'Ivoire, Ghana and Nigeria) over the last decade (Fig. 1.2). Nevertheless, in Ghana national production has increased due to the area under cocoa cultivation expanding at the expense of crop lands and natural forests (Ajagun et al., 2021; Forestry Commission, 2010).

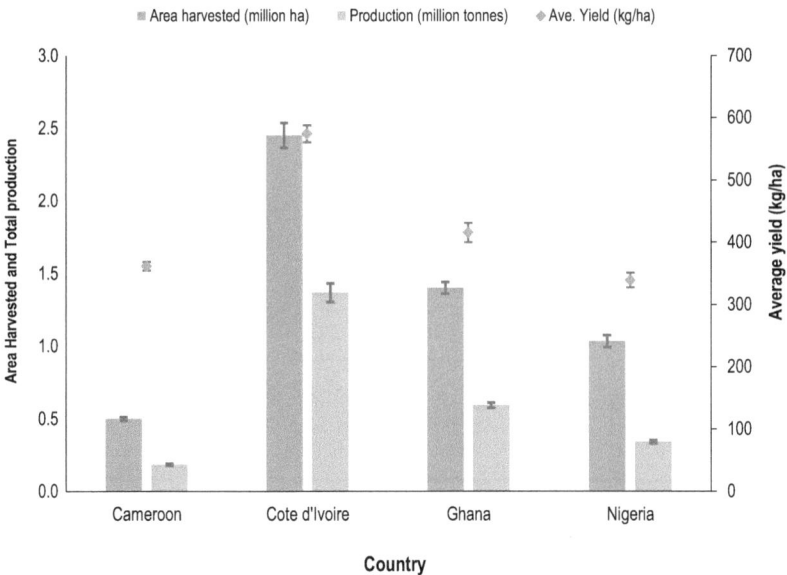

Fig. 1.2 Area harvested, total production and average yields (error bars indicate SE of the mean) of cocoa, 2010–2020 (FAOSTAT, 2021)

Cocoa smallholders have used forest areas as land banks to take advantage of the available nutrients stored in the organic-rich forest soils of newly cleared areas to compensate for falling yields on old cocoa farms (Asare & Ræbild, 2016; Gockowski et al., 2013; Ruf & Zadi, 1998). This is exemplified for Ghana in Fig. 1.3, which shows the annual gross tree cover loss between 2010 and 2020 at a >30% canopy density cover threshold. Currently, about 80% of the Upper Guinean Forest in West Africa has been lost because of cocoa production combined with other land use conversion, such as large-scale surface mining and increasing urbanization (Asare, 2019).

Various studies predict that the West African cocoa belt will experience longer dry seasons and increases in temperatures by 2050 (Läderach et al., 2013; Schroth et al., 2016). This is aggravated by other factors, such as decreased soil moisture and a build-up of pests and diseases, leading to a decrease in land suitable for cocoa cultivation and forcing farmers to expand into forested areas, causing further forest degradation

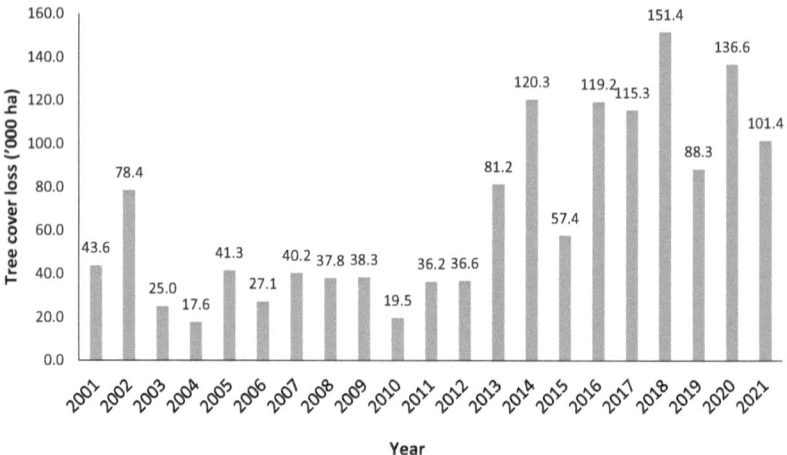

Fig. 1.3 Estimated gross tree cover loss at >30% ('000 ha) canopy density cover threshold in Ghana, 2001–2021 (www.globalforestwatch.org)

(Ruf et al., 2015). It is projected that climate change will have a significant negative impact on many agricultural commodities and communities, including smallholders with a limited capacity to adapt to adverse shocks, further exacerbating global poverty and food insecurity (Howden et al., 2007; Morton, 2007). Thus, both mitigation efforts to reduce greenhouse gas (GHG) emissions and adaptation measures to sustain crop yields are important.

Projected climate change scenarios in West Africa predict a marginal decrease in rainfall along the coastal cocoa-growing areas in countries like Liberia, Côte d'Ivoire and Sierra Leone by 2050 (Läderach et al., 2013; Schroth et al., 2016). Also, longer dry periods are projected in the cocoa belt, with Côte d'Ivoire expecting a gradual drying, and a projected decrease in precipitation in areas beyond Ghana's transition zone (Läderach et al., 2013). The driest parts of Côte d'Ivoire, Ghana, Sierra Leone and Liberia are projected to experience further increases in drought from now toward 2050, which will also affect the forest zones of all cocoa-growing areas along the West African cocoa belt (Schroth et al., 2016). Further, it is argued that changes in the timing of the wet and dry seasons will impact both the cropping patterns of mature cocoa and the successful establishment of cocoa seedlings (Black et al., 2020).

For the West African region, including Ghana, results obtained by the Coupled Model Intercomparison Project Phase 6 (CMIP6) for the high carbon emissions scenario SSP585 indicate that for 2070–2099 compared to the 1985–2014 reference period, overall precipitation in the wet and dry seasons will change very little, but the number of wet days in the rainy season will decrease, implying that the amount of rainfall per rainy day will increase. In the dry season, the number of wet days and the amount of rain per rainy day may increase a little, but these changes are not significant (Wainwright et al., 2021). In the wet season, the mean length of wet spells is expected to decline. Results for the dry season are not significant but indicate a slight decline in the mean length of wet spells. Overall, the simulation results thus indicate that fluctuations tend to become more frequent and more extreme (Wainwright et al., 2021).

Predicting future climatic conditions in West Africa has been difficult, and climate change impacts on cocoa cultivation keep being manifested in different ways (Bunn et al., 2019). For instance, Läderach et al. (2013) and Schroth et al. (2016) predict a gradual decrease in climate suitability for cocoa cultivation in West Africa in the coming decades, with more favorable conditions for cocoa cultivation in the southern cocoa belt compared to the north, which represents the transition zone to the savannah. Modeling work has also shown a likely increase in the annual mean temperature of 2.1 °C by 2050.

1.3.1   Climate-Smart Agriculture

The projected decrease in cocoa-growing suitability in West Africa may be further aggravated by deforestation and soil degradation, but on the other hand, it can be buffered by potential adaptive innovations co-developed by farmers, development partners and scientists, and adopted by rural cocoa-growing communities. Several management strategies have been suggested for simultaneously achieving adaptation and mitigation benefits at the plot, farm and landscape levels. For example, soil conservation practices and the use of conservation agriculture, such as the incorporation of crop residues and cover crops, use of composts, and minimum tillage to increase organic carbon in soils and improve soil moisture through mulching, can all maintain soil fertility and reduce erosion during extreme weather events (Delgado et al., 2011; Hobbs, 2007).

Compared to full-sun cocoa systems, agroforestry leads to increased soil carbon stocks and above-ground biomass, provides shade for protection of the cocoa crop against rising temperatures, diversifies farmers' incomes and helps reduce financial risk (e.g., Matocha et al., 2012; Verchot et al., 2007). Practices like agroforestry that address both adaptation and mitigation goals are referred to as 'climate-smart.' In addition to agroforestry, other climate-smart practices include conservation agriculture, sustainable agriculture, evergreen agriculture, silvopastoral systems, sustainable land management and best-management practices (FAO, 2010; Garrity et al., 2010; Hobbs, 2007; McNeely & Scherr, 2003; Vaast et al., 2016; Vaast & Somarriba, 2014).

FAO (2013) refers to climate-smart agriculture (CSA) as agricultural production with the aim of sustainably increasing productivity and resilience (adaptation), reducing or removing GHG emissions (mitigation) and enhancing the achievement of national food security and development goals. Within the cocoa sector, the concept of CSA is referred to as climate-smart cocoa (CSC) and is defined as a strategy with the potential to sustainably increase yields and incomes while reducing rates of deforestation and forest degradation, as well as enhancing carbon stocks on farms (Asare, 2014; CSCWG, 2011; Vaast et al., 2016). According to Asare et al. (2019), synergies between CSA and the cocoa and forestry sectors include the focus on increasing productivity, the goal of resilience in the face of predicted changes in temperature and rainfall patterns (Läderach et al., 2013), and the mitigation potential from increasing shade tree cover in the cocoa system (Ruf & Zadi, 1998).

1.3.2   Does Cocoa Agroforestry Fit the Bill?

Cocoa agroforestry involves the strategic integration of suitable and valuable non-cocoa tree species and other plants into a cocoa farm at various stages, and management of cocoa farms through the three-dimensional arrangement of trees on the ground and in the canopy (Asare, 2006). Recent studies show positive impacts of this practice on yields (Andres et al., 2018; Asare et al., 2019; Blaser et al., 2018), income (Djokoto et al., 2017) and improvements to environmental integrity (Asare et al., 2014). The role of cocoa agroforestry is influenced by the composition and structural pattern of the agroforestry system, which to a large extent is the result of farmers' decisions to retain tree species according to their perceived values (Abdulai et al., 2018b; Graefe et al., 2017; Smith

Dumont et al., 2014). Food crops, fruit and timber trees in cocoa agroforestry systems are used to give both temporary and permanent shade to the young and mature cocoa plants (Asare & David, 2011). As climate change impacts increase in severity, it becomes more urgent to consider ways to reduce the negative impacts. This could include planting of shade trees since increases in the degree of shade are hypothesized to decrease effects of climatic stress. It is necessary to select shade trees according to farmer preference (Asare, 2005). Furthermore, the selected trees should be compatible with cocoa (Sauvadet et al., 2020) in terms of avoidance of disease and pest attacks (Asitoakor et al., 2022), minimal competition for resources such as nutrients and water (Abdulai et al., 2018a) while providing improvements of the microclimate (Graefe et al., 2017; Sauvadet et al., 2020; Smith Dumont et al., 2014).

Integrating diverse non-cocoa crops potentially provides food for farmers due to higher cropping intensities and diversities. According to Djokoto et al. (2017), farm households engaged in cocoa agroforestry benefit by producing food crops consumed by the household and earn additional incomes from the sale of food produce, including plantains, yams, fruit, honey and vegetables. Recent research (Andres et al., 2018; Asare et al., 2019; Blaser et al., 2018) has shown that cocoa yields reach a maximum at a shade cover of 30–50% due to the efficient use of nutrients and moisture (Isaac et al., 2007) and the reduction in the incidence of pests and diseases (Andres et al., 2018; Asitoakor et al., 2022).

In addition, cocoa agroforestry practices can increase on-farm carbon stocks (Afele et al., 2021) and thereby serve as a potential source of additional household income from carbon credits. The drive for agroforestry can trigger tree-planting practices by farmers in line with ongoing global initiatives like the United Nations' program on Reducing Emissions from Deforestation and forest Degradation (REDD), the Voluntary Carbon Market, the Cocoa and Forest Initiative, and the chocolate industry Cocoa Certification Initiative. These initiatives develop various policies on payment mechanisms to reward farmers for their environmental stewardship. By bundling several sources of revenues, including cocoa, timber, fuelwood, fruits and non-timber products, together with local and international payments for environmental services (carbon sequestration, biodiversity conservation, etc.), cocoa agroforestry can be made attractive and economically rewarding for both young and older farmers. However, some authors (Akrofi-Atitianti et al., 2018; Asare et al., 2014) have raised questions regarding issues of equality concerning how cocoa farmers can be fairly rewarded by carbon credits and other payment schemes, and how doing so could improve adoption. There is not yet sufficient clarity on key issues concerning the development of a fair and transparent benefit-sharing scheme or a definition of carbon rights.

Another important factor to consider is that, when land and tree tenure are not aligned with farmers' indigenous practices, undesirable consequences may result. For example, naturally occurring shade trees in cocoa landscapes have been designated as timber concessions without consideration of their shade function. In Ghana, the Concession Act No. 124, 1962, section 16 (4) states that 'All rights with respect to timber trees on any land is vested in the president in trust for the stools concerned.' This means that naturally occurring timber trees are in principle 'owned' by the community held in trust by the chief and vested in the president. This has led to cocoa farmers removing naturally occurring shade trees by, for example, ringbarking (making deep rings around trunks) or setting fires at the base of trees to avoid damages to their cocoa farms from lumbering (Asare & Prah, 2011). Furthermore, due to customary rights, possession of valuable timber trees, such as timber trees serving as shade trees on cocoa farms, can generate challenges when these trees need to be protected from powerful timber concessionaires or there is a need to negotiate compensation when such trees are harvested by the state (Asare & Ræbild, 2016). This is particularly the case for women, who generally have more insecure land and tree rights than men.

If farmers' tree tenure is ensured, it will become attractive for farmers to nurture naturally occurring shade trees, including shade trees that could eventually be harvested for timber. This could offer significant economic benefits to both farmers and the Ghanaian economy by improving household livelihoods and foreign exchange earnings through timber sales, while providing an environmental resource that helps to ensure a healthy and productive cocoa farm and provide ecological services. Shade trees suitable for timber on farms could also serve as collateral for future retirement benefits (e.g., as a pension scheme) to promote cocoa farmers' investment in trees on farmland.

Projections concerning the reduced suitability for cocoa production in some parts of the West African cocoa-growing belt call for a concerted effort to demonstrate how to combine adaptive strategies, resilience building in farming systems and mitigation practices. According to Harvey et al. (2014), there are substantial opportunities to simultaneously pursue adaptation and mitigation goals in tropical agriculture and adopt integrated landscape approaches that contribute to climate change goals for food security and ecosystem service provision. To ensure the sustainability of cocoa agroforestry, it is important to move beyond a narrow concern with climate change adaptation and mitigation and consider broader social and institutional challenges (Boadi et al., 2022).

1.4   Historical Cocoa Yields and Climate

We now continue with an analysis of historical cocoa production across the last six decades (1960–2020) and its relationship with climate. We examine the development of climate both nationally and regionally to establish a foundation for understanding the climate risks that cocoa farmers must overcome to ensure future sustainable cocoa production in Ghana. To this end, we draw on a range of data sources and describe patterns using descriptive statistics. Formal prediction of future cocoa production is beyond the scope of the book.

1.4.1   Climate and Weather

Historical data (1960–2020) on daily minimum and maximum temperatures and daily precipitation were obtained from the Ghana Meteorological Agency's synoptic stations.[1] Based on these data, mean minimum and maximum temperatures and accumulated precipitation were calculated for monthly and seasonal periods and are used in the analysis. Country-level spatially averaged time series based on gridded historical

---

[1] Data on minimum and maximum temperatures and precipitation were provided by the Ghana Meteorological Agency and originate from a total of 16 synoptic stations, from Axim in the Western Region at 4° 52′ N, 2° 14′ W in the south to Navrongo in the Upper East Region at 10° 53′ N, 1° 5′ W in the far north and from Ada in the Greater Accra Region at 5° 47′ N, 0° 38′ E in the east to Bole in the Savannah Region at 9° 2′ N, 2° 29′ W in the west. Not all stations provide data on all three variables, but from the 1960s data from 10 to 13 stations are available, and from 1981 up to 16 stations have provided data. However, in the years 2013–2020, data are only available from 11 to 15 stations. More specifically, daily minimum temperature data from four stations cover the years 1960–2020 and, except for a few years with missing data, nine stations cover the years 1981–2020. For daily maximum temperature, six stations cover the years 1960–2020 (few exceptions) and 12 stations cover almost all the years 1981–2020. Finally, for precipitation, the years 1960–2020 and 1981–2020 are in most cases covered by three and nine stations, respectively.

data (1960–2020) on monthly mean minimum and maximum temperatures and precipitation were used as a supplement since data from the synoptic stations were not always available.[2]

Previous studies have indicated that the amount and temporal distribution of precipitation is important for cocoa (Anim-Kwapong & Frimpong, 2008). In addition to considering overall trends in precipitation, it is particularly relevant to focus on the drier seasons of the year. Moreover, since both decreasing precipitation and increasing temperatures can cause drought stress, both variables deserve attention. The driest months of the year (in the cocoa-growing areas) are December–February, and the country-level spatially averaged data indicate that, while the mean maximum temperatures in December–February remained stationary with a mean below 34 °C until the early 1990s, an increase to around 34.5 °C on average has been observed since then.

Between 1960 and 1980, country-level annual precipitation showed a declining trend (approx. 1300 mm → 1150 mm) with noticeable year-to-year variation (SD = 138 mm), but since 1980 there has been no clear trend, and the annual variation has been slightly lower (SD = 108 mm). The temporal pattern observed for precipitation in December–February mirrors that of the annual precipitation and shows a decreasing trend from 1960 to 1980, but after then it appears stationary. Particularly dry were December–February in 1981/1982, 1990/1991, 1993/1994 and 2014/2015, with estimated precipitation less than 20 mm.

In Fig. 1.4, the historical development in decadal mean values of monthly precipitation and minimum and maximum temperatures is illustrated for four meteorological stations in the Ashanti, Bono, Eastern and Western Regions (see Fig. 1.1 for the location of these regions) and for the decades 1970–1979 (lighter colors) and 2010–2019 (darker colors). Consistent increases in monthly mean minimum and maximum temperatures (0.5–2 °C) can be observed for all stations and all months of the year, and the increases are typically greatest for maximum temperatures in the dry months of December–January. There is also a slight rainfall decrease in February–July, whereas rainfall has remained approximately unchanged or increased somewhat in September–October. Being

---

[2] National-level spatial average time series based on gridded historical data (1901–2021) on monthly mean minimum and maximum temperatures and precipitation were obtained from the Climatic Research Unit (CRU) of the University of East Anglia (CRU-TS, v. 4.06; https://crudata.uea.ac.uk/cru/data/hrg/; Harris et al., 2020).

located at the coast, Axim (4° 52′ N, 2° 14′ W) sets itself apart from the other stations by having particularly high rainfall in May and June of both decades (>200 mm/month on average). The overall bimodal rainfall pattern with a longer dry season in December–February and a shorter dry season around August is consistent across the stations and over the forty years. The minor wet season starts at all the stations in September, except in Axim where it starts in October.

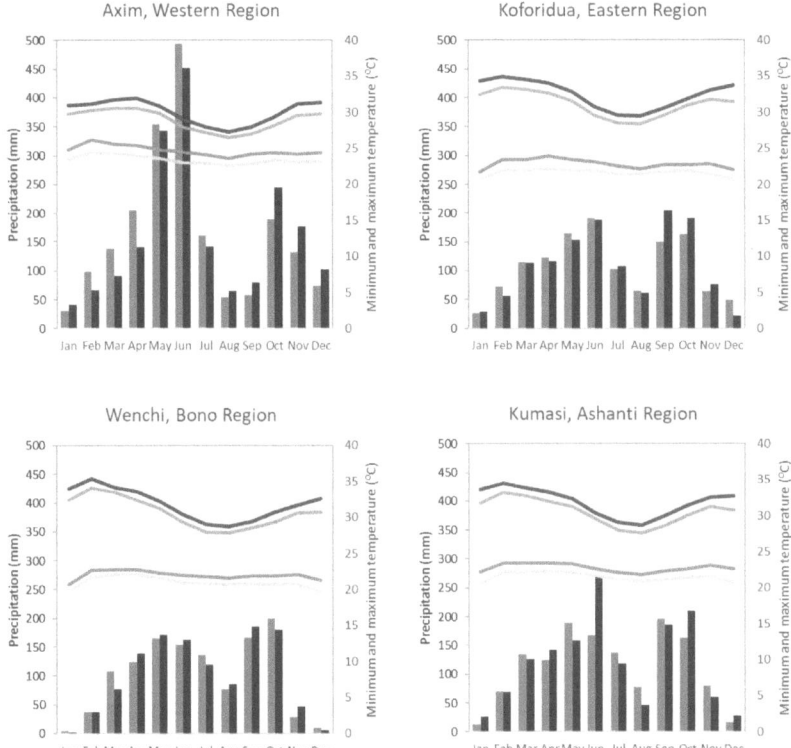

Fig. 1.4 Mean monthly minimum and maximum temperatures (lines) and monthly precipitation (bars), calculated for four meteorological stations in the Ashanti, Bono, Eastern and Western Regions in 1970–1979 (lighter-colored lines and bars) and 2010–2019 (darker-colored lines and bars) (Data courtesy of GMet, Ghana Meteorological Agency)

The diagrams in Fig. 1.4 hide the fact that both the total amount of precipitation and its distribution across the year vary considerably from year to year. To illustrate this, Fig. 1.5 (top panel) shows the monthly precipitation at Kumasi (Ashanti Region). Here, December–February are always fairly dry, March is usually dry but not always, June is usually wet, August is usually dry, and September–October are sometimes wet. However, a severe drought arises when the weather remains dry for a number of consecutive months. Therefore, Fig. 1.5 (bottom panel) also shows cumulative precipitation, calculated for six-month periods. Here, it becomes clear that the cumulative precipitation is mostly low in January–April, whereas in June–November, the cumulative values are high in most years. The most conspicuous exceptions are 1982 and 1983, which is consistent with the incidence of large bushfires, particularly in 1983 linked to El Niño. A few years with particularly high cumulative values in July–November occurred in 1963, 1966 and 1968.

Across the cocoa belt, the year-to-year variation of mean minimum and maximum temperatures in December–February are mostly in synchrony. The mean maximum temperatures (Fig. 1.6) are highest toward the north and east (33–35 °C) and lowest toward the southwest (30–32 °C). The mean regional precipitation in December–February varied somewhat (10–41 mm) within the cocoa belt, with an overall mean of 19 mm for the Bono Region and 35–41 mm for the Eastern, Western and Ashanti Regions. Across the 61 years covered by the data, the mean minimum and maximum temperatures for December–February generally increased by about 0.1–0.4 °C per decade on average, whereas the precipitation in the same three months decreased by 0.7–2.2 mm per decade. On average, a slight increase in aridity can therefore be observed in the three driest months of the year. Moreover, in agreement with the development of the country-level temperature average, it appears that the regional mean maximum temperatures increased slowly or remained almost stationary until around 1990 and then started increasing more rapidly (Fig. 1.6).

1.4.2   *Production and Yields*

Historical production data were obtained from the Ghana Cocoa Board (Cocobod), which regulates the pricing, purchasing, marketing and exportation of cocoa beans in Ghana. The data covers national (1960/1961–2019/2020), regional (1960/1961–2019/2020) and district

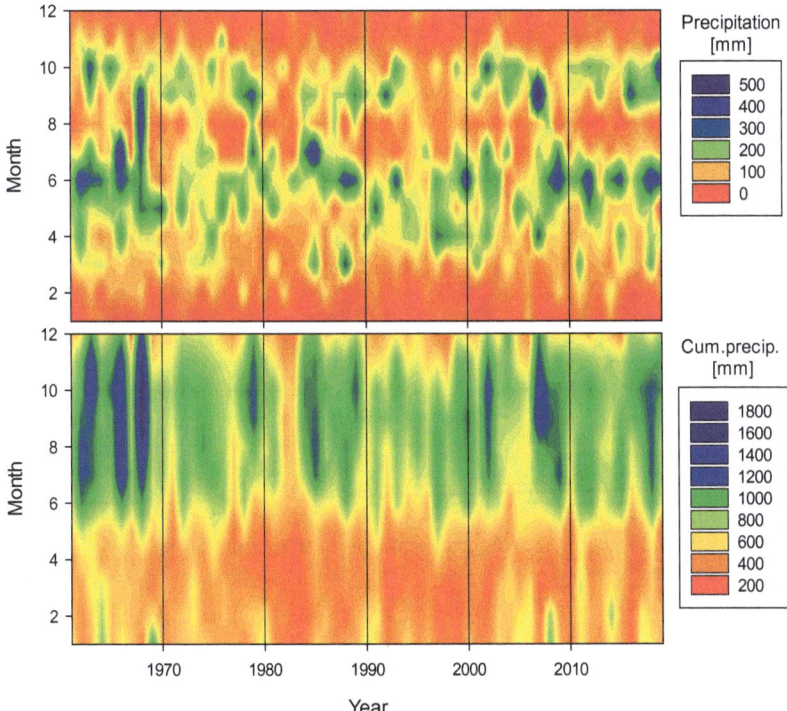

Fig. 1.5 Precipitation in Kumasi (6° 40′ N, 1° 37′ W), Ashanti Region, 1960–2020. Top: monthly precipitation; Bottom: cumulative precipitation calculated for a six-month period, including the current and the five preceding months (Data courtesy of GMet, Ghana Meteorological Agency)

(2000/2001–2014/2015) levels.[3] Historical data on the total area harvested and on average national yields (1961–2020) were obtained from FAOSTAT.[4]

According to Cocobod, the total annual production of cocoa beans was around 400,000 tons from 1961 to the mid-1970s. From 1976, a

---

[3] Regional and national production data (1947/1948–2019/2020) are available at the Ghana Cocoa Board's (Cocobod) webpage (https://cocobod.gh/cocoa-purchases).

[4] FAOSTAT, Statistics Division of the Food and Agriculture Organization of the United Nations (https://www.fao.org/faostat/en/#home).

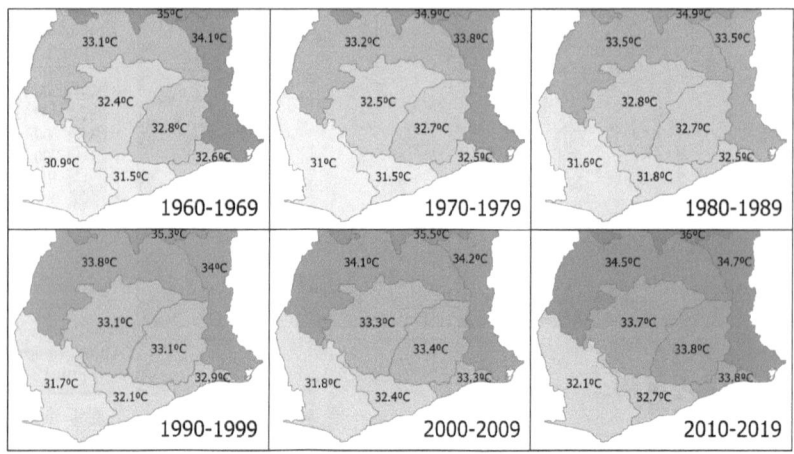

Fig. 1.6 Mean maximum temperatures for December–February in the cocoa belt of Ghana interpolated using data from between five (1960–1969) and fifteen (2010–2019) meteorological stations (Data courtesy of GMet, Ghana Meteorological Agency)

marked decline set in, and a historical minimum of 159,000 tons was reached in 1983/1984, consistent with the droughts experienced in 1982 and 1983. The low production in these years is at least to some extent also affected by cocoa being sold illegally via Côte d'Ivoire and Togo due to the low value of the Ghana Cedi at the time. In the following twenty years, production slowly increased, and from 2004, a harvest of about 700,000 tons was reached. In the years after 2012, production reached a level of 800–900,000 tons annually. According to Cocobod, production peaked in 2010/2011 and again in 2020/2021 at slightly above a million tons.

The regional development in production is illustrated in Fig. 1.7. Until around 1970, the annual production was 50–100% of the historical maxima in the Ashanti, Brong Ahafo, Central, Eastern and Volta Regions, but in the following years, production in these regions declined toward a historical minimum in the early 1980s and then increased again, first slowly in the 1980s and 1990s, and then more rapidly from around 2000. By contrast, in the Western Region annual production was only 20–40,000 tons until the mid-1980s, but it then increased to about ten times as much by 2010 through the establishment of cocoa farms in the virgin forests of this region (Ruf & Schroth, 2004; Ruf et al., 2015). In fact, since the middle of the 1990s, 50–60% of Ghana's annual cocoa production has taken place in the Western Region. In the Volta Region production decreased from 20–30,000 tons annually in the 1960s to 1–4,000 tons annually between 1980 and 2010. Only in the most recent decade has it increased slightly again to 5–8,000 tons/year. The westward shift in cocoa production was partly a consequence of the spreading of the Cocoa Swollen Shoot Virus (CSSV), which started in the 1930s and ultimately led to the abandonment of huge areas of plantations in the east and encouraged cocoa farmers to migrate and establish new plantations in the west (Danquah, 2003).

Aside from the different long-term developments in cocoa production within the regions, clear similarities in the year-to-year short-term fluctuations can be observed across the regions from 1960/1961 to 2019/2020. A year of comparatively high production in one region was typically also a good year in other regions, and coefficients of correlation

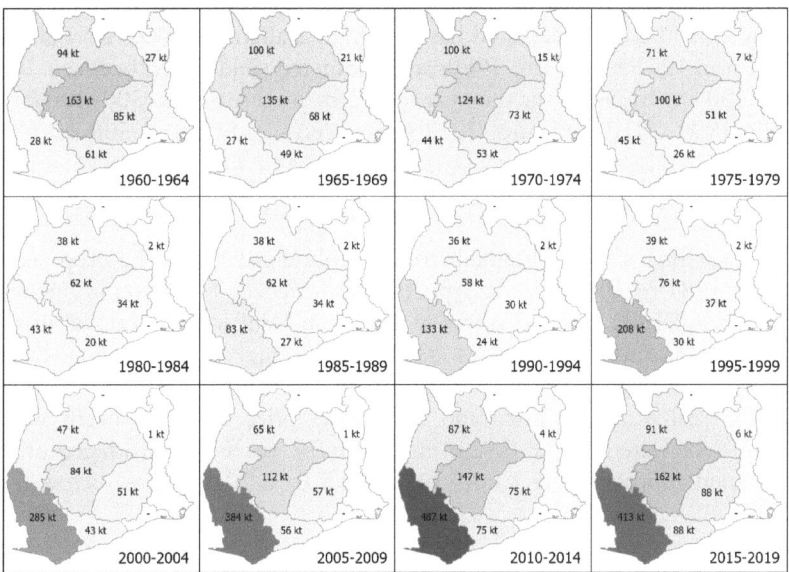

Fig. 1.7 Average production of cocoa beans in the cocoa belt of Ghana, calculated for five-year periods (1960–1964, …, 2015–2019) by region based on Cocobod purchases. Unit: kilo tons, kt = 1,000,000 kg

between detrended annual production figures for different regions typically exceeded 0.6. This indicates that, at least in the short term, the production is affected by one or more common factors, of which the weather could be an important one. For the Volta Region, the similarity with other regions is only clear in the 1960s, before its production began declining to its current low level.

According to FAOSTAT, the total area harvested in Ghana first peaked in 1964, where it reached 1.85 million ha. In the following 24 years, the area gradually declined and reached a minimum of about 0.7 million ha in 1989–1994. The long decline was due to increasing cocoa taxation, the Ghanaian government's main source of income, an over-valued Ghana Cedi and high inflation rates, which made cocoa cultivation a poor business for farmers (Kolavalli, 2019). The decline was further exacerbated by low precipitation, widespread disease and bushfires in the early 1980s. In the decade following 1994, as a result of increasing world market prices and renewed political efforts to strengthen and promote the Ghanaian cocoa sector both domestically and in global markets, the areas under cocoa increased and peaked at the historical maximum of 2.0 million ha. Since then the area harvested has seemingly stabilized at 1.6–1.7 million ha. From 1961 to the late 1980s, the estimated yield varied between 200 and 300 kg/ha, and over the next few years, it increased to 300–400 kg/ha in 1990–2010, and since 2012 it has been slightly above 500 kg/ha.

1.4.3   Producer Prices

Producer prices were obtained from Cocobod and cover the crop years 1980/1981 to 2017/2018. Export prices for cocoa beans were obtained from FAOSTAT, and real prices in Ghanaian Cedi were calculated using annual consumer price indices and annual average exchange rates from the World Bank.[5]

The real producer price and the export price of cocoa beans were highly correlated ($r = 0.92$), and in most years the producer price remained approximately 1,000–2,000 Cedi (real 2010 prices) below the

---

[5] Annual average consumer price indices (1964–2017) and annual average exchange rates (Ghanaian Cedi per US dollar, 1964–2017) were obtained from the World Bank (World Development Indicators, https://data.worldbank.org/). Export prices for cocoa beans (1961–2020) were obtained from FAOSTAT (https://www.fao.org/faostat/en/#home).

export price. In the second half of the 1980s, producer prices were 40–50% of the mean export price, but during the 1990s, Cocobod gradually increased the producer price to a level corresponding to about 60–70% of the mean export price. This reflects Cocobod's partial liberalization of the cocoa-bean market and a price policy that enables farmers to plan under the assumption that they will not experience a drop in price.

While short-term fluctuations in production may be related to variations in the annual weather conditions, variations in the longer term could partly be explained by changing prices. Hence, it emerges that across the 37 years covered by the dataset, the real producer price was highly correlated with annual production ($r = 0.82$), the area harvested ($r = 0.70$) and the yield of beans per hectare ($r = 0.67$).

1.4.4 Association Between Production and Climate Variables

To examine correlations between year-to-year fluctuations in climate variables and in regional or national cocoa production, the time series were first detrended using polynomial trend models. This eliminated long-term variation not related to the weather but also helped stabilizing the variance. Next, year-to-year changes were calculated to reduce the serial correlation. The analysis was carried out for both the interpolated regional climate series and the national-level spatially averaged series. Correlation patterns were mostly similar but results for precipitation turned out to be more consistent for the national average series, and therefore these are used here.

Across all six regions in the dataset (Ashanti, Brong Ahafo, Eastern, Central, Western and Volta), the year-to-year changes in production were correlated with changes in dry-season precipitation, especially in November. Hence, except for the Volta Region, which was hit very hard by bushfires in the early 1980s and has had very low production levels since then (cf. Fig. 1.7), the correlation of changes in regional production with changes in November's precipitation generally exceeded 0.30 ($0.30 < r < 0.43$, $p < 0.05$). Similarly, the positive correlation with precipitation in the three-month period December–February was in most cases significant at the 5% level.

For temperature, particularly high and significant ($p < 0.05$, except for the Volta Region) negative correlations were observed between year-to-year changes in regional production and changes in July's maximum temperature. Correlations of year-to-year changes in production with changes in temperature for June, August and September were also negative but not always significant at the 5% level. Correlations with changes in maximum temperature for the three-month periods June–August and July–September were similar to those observed for July. Correlations estimated for the three-month period December–February were also negative but slightly weaker than those observed for July and June–August. For the months June–September, correlations estimated for minimum temperatures turned out to be similar to those obtained for the maximum temperatures, but for the three-month period December–February the correlations were not significant.

1.4.5   Consistency of the Correlation Patterns

Previous studies have observed that increasing temperatures and reduced precipitation tend to reduce cocoa production (e.g., Ofori-Boateng & Insah, 2014). In agreement with the assumption that cocoa production is affected the most at times of the year when precipitation is low or absent for an extended period of time, we found significant and positive correlations between year-to-year variation in production and precipitation in and around the major dry season, particularly in the month of November, as also reported by Asante et al. (2022), and in the three-month period December–February. In addition, and in agreement with the possible incidence of heat stress, we found significant and negative correlations with variations in maximum and minimum temperatures in July and in the three-month periods June–August and July–September. In agreement with the results of Schroth et al. (2016), clear negative correlations were also observed with maximum temperature in the December–February period. Since maximum temperature and precipitation in December–February are negatively correlated, the negative correlation of production with maximum temperature may essentially reflect the same response as the positive correlation with precipitation. Furthermore, the correlation between minimum temperatures and precipitation in the dry season is much weaker than for maximum temperatures and precipitation, thus presumably explaining why the correlation between production and minimum temperatures in December–February was not significant.

In agreement with the trend observed over the last sixty years (cf. Fig. 1.6), climate projections consistently indicate that increasing temperatures can be expected, but for Ghana, the forecasted changes in precipitation patterns are less clear and likely to be small (e.g., Wainwright et al.,

2021), including during the dry season, and in agreement with historical patterns (cf. Figs. 1.4 and 1.5). Thus, increasing temperatures and the associated evaporative demand, leading to water stress and possibly also to heat stress, may be more direct causes of a likely decline in cocoa production than the precipitation as such.

1.5   OVERVIEW OF CHAPTERS

This chapter's analysis of historical Ghanaian cocoa yields and climate data has highlighted changes in climate and how these historically have influenced yields. Building on previous cocoa agroforestry research that has largely focused on the health of cocoa as an export crop, the chapters that follow will contribute important findings on the complex of plant species and their environmental and societal attributes, the social and institutional contexts within which agroforestry practices are introduced across a climate gradient, and the impacts of climate change on the socio-biophysical bases of cocoa systems.

Chapter 2 investigates how full-sun cocoa systems versus cocoa agroforestry systems perform under changing climate and soil conditions. The chapter first discusses the impact of soil moisture as well as environmental stress, specifically drought and heat, on cocoa plants. It then examines how the impact of these factors on cocoa plant growth and yield are changed by shade. This chapter thus contributes to the still limited number of studies on the impact of increased heat and drought on the physiology of the cocoa tree and discusses how and if shade trees may reduce these effects. The chapter shows that shade generally reduces stress effects on cocoa plants, but that when temperatures and drought reach a certain level, shade only partially compensates for the negative effects. It draws on data generated from two experiments that were set up to investigate the effects of drought and elevated temperatures on cocoa, one with heat and shade using seedlings, and the other with drought and shade using a mature cocoa stand. The focus is on the physiological responses of cocoa trees to environmental stress and different levels of shade, which is a crucial first step in understanding the different factors that influence the potential of agroforestry as a means of climate change adaptation in cocoa farming.

Chapter 3 focuses on the species-specific effects of shade trees on cocoa, a topic that is in its infancy in cocoa agroforestry research. The chapter examines how various shade tree species impact the general health and productivity of the cocoa tree, including soil fertility and pests and disease infestation. Overall, results confirm that, under low input conditions (of fertilizer and pesticides), shade increases yields compared to full-sun systems. Furthermore, the chapter shows that shade tree species may have different effects on yields and the occurrence of pests and diseases, offering opportunities for improved management by strategically selecting tree species appropriate to the local context. The chapter is based on a four-year on-farm study of eight different agroforestry shade tree species and their effects on cocoa trees and their yields, as well as the prevalence of pests and diseases that damage cocoa pods. This chapter thus provides data to substantiate the overall claim of the book that site-specific and tree-specific biophysical knowledge is important to fully realize the potentials of agroforestry.

Chapter 4 moves from a biophysical to a socioeconomic focus and discusses the social challenges and opportunities linked to agroforestry from the perspective of the cocoa farmers. It describes the different monetary and non-monetary values farmers obtain from shade trees on cocoa farms, but argues that social barriers and institutional factors, such as land and tree rights, can prevent cocoa farmers from engaging in longer-term agroforestry practices and thereby from benefitting from the opportunities shade trees present. Based on focus-group discussions and in-depth interviews with cocoa farmers in twelve different cocoa-growing communities, the chapter shows that to realize the opportunities of agroforestry, it is necessary for farmers to navigate a complex socioeconomic landscape, including issues related to access, rights, ethnicity, migration, gender and other institutional and cross-cutting factors. It focuses specifically on the lesser-known benefits of shaded cocoa fields in the form of mushrooms and snails, and on challenges such as the role of chiefs, who are found to wield unchallenged power over land possessions, as well as the influence of mining activities that literally remove the foundation for investments in trees and cocoa. These and other very tangible challenges to cocoa agroforestry are the result of a mixture of historical land policies and contemporary opportunities for a more secure income.

Chapter 5 looks at another important socioeconomic factor, namely the household economics of cocoa agroforestry. To understand farmers' decision-making in relation to agroforestry, this chapter examines the costs and benefits of cocoa agroforestry along a climate gradient. It is based on household surveys in three different climate zones and is thus able to examine how different climates impact the costs and benefits of agroforestry. The chapter argues that cocoa agroforestry can be more profitable than monocrop systems when combined with income from shade trees, and that agroforestry farms with fruit trees are more profitable and more competitive across all three climate zones. Moreover, cocoa farmers attain a consistent and sustained profit from their cocoa plots if they implement a more tree species-diverse cultivation system that includes fruit trees suitable to local needs and conditions. Importantly, hired labor costs were lower the higher the tree species diversity because suitably shaded cocoa requires less intensive care than full-sun systems. This is important not just for reducing costs, but also because reduced labor inputs free up labor for both on-farm and off-farm diversification activities. The chapter thus further illustrates the complexity of integrating trees into cocoa farming systems, and the importance of not just looking at shade levels, as is commonly done in cocoa agroforestry research, but also paying attention to shade tree species diversity.

In the concluding Chapter 6, the findings presented in the different chapters are brought together in making the overall argument of the book, namely that agroforestry can only successfully address climate change impacts on cocoa farming if site-specific biophysical and socioeconomic factors are considered. The chapter also summarizes and discusses the overall policy and practice implications for cocoa farming that arise from the findings of the book. Finally, the chapter points to important new directions suggested by this study for further and broader research on the potential of climate change adaptation, which considers social and ecological factors in the larger context of agriculture vulnerable to climate change.

REFERENCES

Abdulai, I., Jassogne, L., Graefe, S., Asare, R., Van Asten, P., Läderach, P., & Vaast, P. (2018a). Characterization of cocoa production, income diversification and shade tree management along a climate gradient in Ghana. *PLoS*

ONE, 13(4), Article e0195777. https://doi.org/10.1371/journal.pone.019 5777

Abdulai, I., Vaast, P., Hoffmann, M. P., Asare, R., Jassogne, L., Van Asten, P., Rötter, R. P., & Graefe, S. (2018b). Cocoa agroforestry is less resilient to sub-optimal and extreme climate than cocoa in full sun. *Global Change Biology, 24*(1), 273–286. https://doi.org/10.1111/gcb.13885

Afele, J. T., Dawoe, E., Abunyewa, A. A., Afari-Sefa, V., & Asare, R. (2021). Carbon storage in cocoa growing systems across different agroecological zones in Ghana. *Pelita Perkebunan, 37*(1), 32–49. https://doi.org/10.22302/iccri.jur.pelitaperkebunan.v37i1.395

Ajagun, E. O., Ashiagbor, G., Asante, W. A., Gyampoh, B. A., Obirikorang, K. A., & Acheampong, E. (2021). *Cocoa eats the food: Expansion of cocoa into food croplands in the Juabeso District, Ghana* (pp. 1–20). Food Security.

Akrofi-Atitianti, F., Speranza, C. I., Bockel, L., & Asare, R. (2018). Assessing climate smart agriculture and its determinants of practice in Ghana: A case of the cocoa production system. *Land, 7*(1), 30. https://doi.org/10.3390/land7010030

Allman, J. (1994). Making mothers: Missionaries, medical officers and women's work in Colonial Asante, 1924–1945. *History Workshop, 38*, 23–47. https://www.jstor.org/stable/4289318

Ameyaw, L. K., Ettl, G. J., Leissle, K., & Anim-Kwapong, G. J. (2018). Cocoa and climate change: Insights from smallholder cocoa producers in Ghana regarding challenges in implementing climate change mitigation strategies. *Forests, 9*(12), 742. https://doi.org/10.3390/f9120742

Andres, C., Blaser, W. J., Dzahini-Obiatey, H. K., Ameyaw, G. A., Domfeh, O. K., Awiagah, M. A., Gattinger, A., Schneider, M., Offei, S. K., & Six, J. (2018). Agroforestry systems can mitigate the severity of cocoa swollen shoot virus disease. *Agriculture, Ecosystems & Environment, 252*, 83–92.

Anim-Kwapong, G. J., & Frimpong, E. B. (2008). Climate change in cocoa production. In W. Agyeman-Bonsu (Ed.), *Ghana climate change impacts, vulnerability and adaptation assessments under The Netherlands Climate Assistance Programme (NCAP)* (pp. 263–298). Environmental Protection Agency.

Asante, P. A., Rahn, E., Zuidema, P. A., Rozendall, D. M. A., van der Baan, M. E. G., Läderach, P., Asare, R., Cryer, N. C., & Anten, N. P. R. (2022). The cocoa yield gap in Ghana: A quantification and an analysis of factors that could narrow the gap. *Agricultural Systems, 201*, 103473. https://doi.org/10.1016/j.agsy.2022.103473

Asare, R. (2005). *Cocoa agroforests in West Africa: A look at activities on preferred trees in the farming systems* (p. 89). Forest & Landscape Denmark (FLD).

Asare, R. (2006). *Learning about neighbour trees in cocoa growing systems—A manual for farmer trainers* (Development and Environment Series 4; 79 pp.). Forest & Landscape Denmark.

Asare, R. (2019, July–December). The nexus between cocoa production and deforestation. In: Ghana, an agricultural exception in West Africa? *GRAIN DE SEL Magazine, 78*, 26–27.

Asare, R. A. (2014). *Understanding and defining climate-smart cocoa: Extension, inputs, yields and farming practices* (Climate-smart cocoa working group). Nature Conservation Research Centre and Forest Trends.

Asare, R., Afari-Sefa, V., Osei-Owusu, Y., & Pabi, O. (2014). Cocoa agroforestry for increasing forest connectivity in a fragmented landscape in Ghana. *Agroforestry Systems, 88*(6), 1143–1156. https://doi.org/10.1007/s10457-014-9688-3

Asare, R., & David, S. (2011). *Good agricultural practices for sustainable cocoa production: A guide for farmer training, manual no. 1: Planting, replanting and tree diversification in cocoa systems* (Development and Environment Series 13–2010; 126 pp.). Forest & Landscape Denmark.

Asare, R., Markussen, B., Asare, R. A., Anim-Kwapong, G., & Ræbild, A. (2019). On-farm cocoa yields increase with canopy cover of shade trees in two agroecological zones in Ghana. *Climate and Development, 11*(5), 435–445. https://doi.org/10.1080/17565529.2018.1442805

Asare, R., & Prah, C. (2011, May 21). Shade trees in the cocoa landscapes: How Ghana can benefit. *Daily Graphic Ghana*, 21.

Asare, R., & Ræbild, A. (2016). Tree diversity and canopy cover in cocoa systems in Ghana. *New Forests, 47*(2), 287–302. https://doi.org/10.1007/s11056-015-9515-3

Asitoakor, B. K., Vaast, P., Ræbild, A., Ravn, H. P., Eziah, V. Y., Owusu, K., Mensah, E. O., & Asare, R. (2022). Selected shade tree species improved cocoa yields in low-input agroforestry systems in Ghana. *Agricultural Systems, 202*, 103476.

Black, E., Pinnington, E., Wainwright, C., Lahive, F., Quaife, T., Allan, R. P., Cook, P., Daymond, A., Hadley, P., McGuire, P. C., & Verhoef, A. (2020). Cocoa plant productivity in West Africa under climate change: A modelling and experimental study. *Environmental Research Letters, 16*(1), 014009.

Blaser, W. J., Oppong, J., Hart, S. P., Landolt, J., Yeboah, E., & Six, J. (2018). Climate-smart sustainable agriculture in low-to-intermediate shade agroforests. *Nature Sustainability, 1*(5), 234–239.

Boadi, S. A., Olwig, M. F., Asare, R., Bosselmann, A. S., & Owusu, K. (2022). The role of innovation in sustainable cocoa cultivation: Moving beyond mitigation and adaptation. In M. Coromaldi & S. Auci (Eds.), *Climate-induced innovation: Mitigation and adaptation to climate change* (pp. 47–80). Springer. https://doi.org/10.1007/978-3-031-01330-0_3

Bunn, C., Fernandez-Kolb, P., Asare, R, & Lundy, M. (2019). *Climate smart cocoa in Ghana: Towards climate resilient production at scale* (CCAFS Info Note). CGIAR Research Program on Climate Change, Agriculture and Food Security (CCAFS). https://hdl.handle.net/10568/103770

CSCWG. (2011). *The case and pathway towards a climate-smart cocoa future for Ghana* (Technical report. Climate Smart Cocoa Working Group). Nature Conservation Research Centre and Forest Trends.

Danquah, F. K. (2003). Sustaining a West African cocoa economy: Agricultural science and the Swollen Shoot contagion in Ghana, 1936–1965. *African Economic History* (31), 43–74.

Delgado, J. A., Groffman, P. M., Nearing, M. A., Goddard, T., Reicosky, D., Lal, R., Kitchen, N. R., Rice, C. W., Towery, D., & Salon, P. (2011). Conservation practices to mitigate and adapt to climate change. *Journal of Soil Water Conservation, 66*, 118A-129A.

Djokoto, J. G., Afari-Sefa, V., & Addo-Quaye, A. (2017). Vegetable diversification in cocoa-based farming systems Ghana. *Agriculture & Food Security, 6*(1), 1–10.

FAO. (2010). *'Climate-smart' agriculture: Policies, practices and financing for food security, adaptation and mitigation.* Food and Agriculture Organization of the United Nations. Retrieved on 15 March 2023, from https://www.fao.org/3/i1881e/i1881e00.pdf

FAO. (2013). *Climate-smart agriculture sourcebook.* Retrieved on March 2023, from http://www.fao.org/docrep/018/i3325e/i3325e.pdf

FAOSTAT. (2021). *Food and Agricultural Organization of the United Nation Statistics Data.* Retrieved on 28 March 2023, from http://www.fao.org/faostat/en/#data/QC

Forestry Commission. (2010). *Revised readiness preparation proposal Ghana.* Submitted to Forest Carbon Partnership Facility (FCPF). Retrieved on 17 March 2023, from https://www.oldwebsite.fcghana.org/assets/file/Publications/Revised%20Ghana%20R-PP_23-Dec-2010%5B2%5D_1_-Version%202.pdf

Garrity, D. P., Akinnifesi, F. K., Ajayi, O. C., Weldesmayat, S. G., Mowo, J. G., Kalinganire, A., Larwanou, M., & Bayala, J. (2010). Evergreen agriculture: A robust approach to sustainable food security in Africa. *Food Security, 2*, 197–214.

Gockowski, J., Afari-Sefa, V., Sarpong, D. B., Osei-Asare, Y. B., & Agyeman, N. F. (2013). Improving the productivity and income of Ghanaian cocoa farmers while maintaining environmental services: What role for certification? *International Journal of Agricultural Sustainability, 11*(4), 331–346.

Graefe, S., Meyer-Sand, L.F., Chauvette, K., Abdulai, I., Jassogne, L., Vaast, P., & Asare, R. (2017). Evaluating farmers' knowledge of shade trees in different cocoa agro-ecological zones in Ghana. *Human Ecology, 45*(3), 321–332. https://doi.org/10.1007/s10745-017-9899-0

Harvey, C. A., Chacón, M., Donatti, C. I., Garen, E., Hannah, L., Andrade, A., Bede, L., Brown, D., Calle, A., Chará, J., Clement, C., Gray, E., Hoang, M. H., Minang, P., Rodríguez, A. M., Seeberg-Elverfeldt, C., Semroc, B., Shames, S., Smukler, S., ... Wollenberg, E. (2014). Climate-smart landscapes: Opportunities and challenges for integrating adaptation and mitigation in tropical agriculture. *Conservation Letters, 7*(2), 77–90. https://doi.org/10.1111/conl.12066

Harris, I., Osborn, T. J., Jones, P., & Lister, D. (2020). Version 4 of the CRU TS monthly high-resolution gridded multivariate climate dataset. *Scientific Data, 7*, 109. https://doi.org/10.1038/s41597-020-0453-3

Hobbs, P. R. (2007). Conservation agriculture: What is it and why is it important for future sustainable food production? *The Journal of Agricultural Science, 145*(2), 127.

Howden, S. M., Soussana, J. F., Tubiello, F. N., Chhetri, N., Dunlop, M., & Meinke, H. (2007). Adapting agriculture to climate change. *Proceedings of the National Academy of Sciences of the USA, 104*, 19691–19696.

IPCC. (2013). *Climate change 2013: The physical science basis. Contribution of working group I to the fifth assessment report of the Intergovernmental Panel on Climate Change* (Stocker, T. F., Qin, D., Plattner, G.-K., Tignor, M., Allen, S. K., Boschung, J., Nauels, A., Xia, Y., Bex, V., & Midgley, P. M., Eds.). Cambridge University Press.

Isaac, M. E., Timmer, V. R., & Quashie-Sam, S. J. (2007). Shade tree effects in an 8-year-old cocoa agroforestry system: Biomass and nutrient diagnosis of Theobroma cacao by vector analysis. *Nutrient Cycling in Agroecosystems, 78*(2), 155–165.

Kolavalli, S. (2019). Developing agricultural value chains. In X. Diao, X. P. Hazell, S. Kolavalli, & D. Resnick (Eds.), *Ghana's economic and agricultural transformation. Past performance and future prospects* (pp. 210–240). IFPRI, Oxford University Press.

Kolavalli, S., & Vigneri, M. (2011). Cocoa in Ghana: Shaping the success of an economy. In P. Chuhan-Pole & M. Angwafo (Eds.), *Yes, Africa can: Success stories from a dynamic continent* (pp. 201–217). World Bank Publication.

Läderach, P., Martinez-Valle, A., Schroth, G., & Castro, N. (2013). Predicting the future climatic suitability for cocoa farming of the world's leading producer countries, Ghana, and Côte d'Ivoire. *Climatic Change, 119*(3), 841–854.

Matocha, J., Schroth, G., Hills, T., & Hole, D. (2012). Integrating climate change adaptation and mitigation through agroforestry and ecosystem conservation. In P. Nair & D. Garrity (Eds.), *Agroforestry—The future of global land use* (Advances in Agroforestry, 9; pp. 105–126). Springer. https://doi.org/10.1007/978-94-007-4676-3_9

McNeely, J. A., & Scherr, S. J. (Eds.). (2003). *EcoAgriculture: Strategies to feed the world and save wild biodiversity*. Island Press.

Morton, J. F. (2007). The impact of climate change on smallholder and subsistence agriculture. *Proceedings of the National Academy of Sciences of the USA, 104*, 19680–19685. https://doi.org/10.1073/pnas.0701855104

Ofori-Boateng, K., & Insah, B. (2014). The impact of climate change on cocoa production in West Africa. *International Journal of Climate Change Strategies and Management, 6*(3). https://www.emerald.com/insight/content/doi/10.1108/IJCCSM-01-2013-0007/full/html

Ruf, F., & Schroth, G. (2004). Chocolate forests and monocultures: A historical review of cocoa growing and its conflicting role in tropical deforestation and forest conservation. In G. Schroth, A. B. da Fonseca, C. A. Harvey, C. Gascon, H. L. Vasconcelos, & A.-M. N. Izac (Eds.), *Agroforestry and biodiversity conservation in tropical landscapes* (pp. 107–134). Island Press.

Ruf, F., Schroth, G., & Doffangui, K. (2015). Climate change, cocoa migrations, and deforestation in West Africa—What does the past tell us about the future? *Sustainability Science, 10*, 101–111. https://doi.org/10.1007/s11625-014-0282-4

Ruf, F., & Zadi, H. (1998). Cocoa: From deforestation to reforestation. Paper from workshop on shade grown cocoa held in Panama, 3/30-4/2. International conference on shade grown cacao, Smithsonian Institution, Washington.

Sadhu, S., Kysia., K., Onyango, L., Zinnes, C., Lord, S., Monnard, A., & Arelleno, I.R. (2020). *NORC final report: Assessing progress in reducing child labor in Cocoa production in cocoa growing areas of Côte d'Ivoire and Ghana* (p. 301). NORC, University of Chicago.

Sauvadet, M., Asare, R., & Isaac, M. E. (2020). Evolutionary distance explains shade tree selection in agroforestry systems. *Agriculture, Ecosystems and Environment, 304*, 107125. https://doi.org/10.1016/j.agee.2020.107125

Schroth, G., Läderach, P., Martinez-Valle, A. I., Bunn, C., & Jassogne, L. (2016). Vulnerability to climate change of cocoa in West Africa: Patterns, opportunities, and limits to adaptation. *Science of the Total Environment, 556*, 231–241. https://doi.org/10.1016/j.scitotenv.2016.03.024

Smith Dumont, E., Gnahoua, G. M., Ohouo, L., Sinclair, F. L., & Vaast, P. (2014). Farmers in Côte d'Ivoire value integrating tree diversity in cocoa for the provision of ecosystem services. *Agroforestry Systems, 88*(6), 1047–1066.

Vaast, P., Harmand, J. M., Rapidel, B., Jagoret, P., & Deheuvels, O. (2016). Coffee and cocoa production in agroforestry—A climate-smart agriculture model. In E. Torquebiau (Eds.), *Climate change and agriculture worldwide*. Springer. https://doi.org/10.1007/978-94-017-7462-8_16

Vaast, P., & Somarriba, E. (2014). Trade-offs between crop intensification and ecosystem services: The role of agroforestry in cocoa cultivation. *Agroforestry Systems, 88*, 947–956. https://doi.org/10.1007/s10457-014-9762-x

Verchot, L. V., Noordwijk, M., Kandji, S., Tomich, T., Ong, C., Albrecht, A., Mackensen, J., Bantilan, C., Anupama, K. V., & Palm, C. (2007). Climate change: Linking adaptation and mitigation through agroforestry. *Mitigation and Adaptation Strategies for Global Change, 12*, 901–918.

Wainwright, C. M., Black, E., & Allan, R. P. (2021). Future changes in wet and dry season characteristics in CMIP5 and CMIP6 simulations. *Journal of Hydrometeorology, 22*, 2329–2357.

Yaro, J. A., Teye, J. K., & Wiggins, S. (2021). *Land and labour relations on cocoa farms in Sefwi, Ghana: Continuity and change* (APRA Working Paper 73). Future Agricultures Consortium.

Open Access This chapter is licensed under the terms of the Creative Commons Attribution 4.0 International License (http://creativecommons.org/licenses/by/4.0/), which permits use, sharing, adaptation, distribution and reproduction in any medium or format, as long as you give appropriate credit to the original author(s) and the source, provide a link to the Creative Commons license and indicate if changes were made.

The images or other third party material in this chapter are included in the chapter's Creative Commons license, unless indicated otherwise in a credit line to the material. If material is not included in the chapter's Creative Commons license and your intended use is not permitted by statutory regulation or exceeds the permitted use, you will need to obtain permission directly from the copyright holder.

CHAPTER 2

Cocoa Under Heat and Drought Stress

Eric Opoku Mensah◉, Philippe Vaast◉, Richard Asare◉,
Christiana A. Amoatey◉, Kwadwo Owusu◉,
Bismark Kwesi Asitoakor◉, and Anders Ræbild◉

Abstract Cocoa (*Theobroma cacao* L.) is an important cash crop in many tropical countries, particularly in West Africa. Heat and drought are both known to affect the physiology of cocoa plants through reduced rates of photosynthesis and transpiration, as well as changed physiological processes such as the functions of photosystems, chlorophyll synthesis, stomatal conductance and expression of heat-shock proteins. This in turn leads to decreased yields and increased risks of mortality under severe heat and drought. To help cocoa plants adapt to climate change, the literature suggests agroforestry as a potential farm management practice. It has been argued that the lack of tree cover in cocoa cultivation systems exposes

E. O. Mensah (✉) · C. A. Amoatey · B. K. Asitoakor
Department of Crop Science, University of Ghana, Accra, Ghana
e-mail: omedjin@gmail.com

C. A. Amoatey
e-mail: camoatey@ug.edu.gh

B. K. Asitoakor
e-mail: bkasitoakor001@st.ug.edu.gh; bka@ign.ku.dk

© The Author(s) 2024
M. F. Olwig et al. (eds.), *Agroforestry as Climate Change Adaptation*,
https://doi.org/10.1007/978-3-031-45635-0_2

the crop to heat and direct solar radiation, thus increasing evapotranspiration and the risk of drought. Drawing on data generated from two on-field studies, this chapter assesses the shade effect on cocoa's physiological responses to drought and heat stress to determine whether shade would be beneficial under climate change scenarios. We conclude that shade improves the physiology of cocoa, but that this may not be sufficient to compensate for the negative effects of high temperatures and severe drought exacerbated by climate change in sub-optimal conditions.

Keywords Cocoa · Heat · Photosynthesis · Shade · Stomatal conductance · Water potential

E. O. Mensah · B. K. Asitoakor · A. Ræbild
Department of Geosciences and Natural Resource Management, University of Copenhagen, Frederiksberg, Denmark
e-mail: are@ign.ku.dk

E. O. Mensah · B. K. Asitoakor
CSIR-Plant Genetic Resources Research Institute, Bunso, Ghana

P. Vaast
UMR Eco & Sols, Centre de Coopération Internationale en Recherche Agronomique Pour Le Développement (CIRAD), Université Montpellier, Montpellier, France

World Agroforestry Center, Nairobi, Kenya

P. Vaast
e-mail: philippe.vaast@cirad.fr

R. Asare
International Institute of Tropical Agriculture (IITA), Accra, Ghana
e-mail: r.asare@cgiar.org

K. Owusu
Department of Geography and Resource Development, University of Ghana, Accra, Ghana
e-mail: kowusu@ug.edu.gh

2.1   Introduction

Cocoa (*Theobroma cacao* L.) is native to South America and belongs to the Malvaceae family (formerly Sterculiaceae). For this species, three main genetic groups are recognized based on physical, sensory quality and associated botanical traits: Forastero, Criollo and Trinitario (Bartley, 2005; Cheesman, 1944). Around 95% of all cocoa production comes from the Forastero and the Trinitario groups, which are high-yielding, more vigorous and less susceptible to pests and diseases than the Criollo group (Loor et al., 2009; Umaharan, 2018).

Cocoa is mostly grown in a narrow belt 20 degrees north and south of the equator with warm and humid tropical climates, regular rains and short dry seasons (Mattayasovszky, 2017). It is mostly planted in smallholder plantations in West Africa, Southeast Asia and Latin America (Lahive et al., 2019). Cocoa plants grow well within a temperature range of 18–32 °C, with regular rainfall of 1000–2500 mm per year and at altitudes as high as 1000 m above sea level (Ameyaw et al., 2018; ICCO, 2020; Wood & Lass, 1992). Under shade, cocoa physiology is changed, and yields may increase (Asare et al., 2017; Tee et al., 2018). Cultivation under full sun without any vegetation cover increases the risk of exposing the crop to the negative consequences of high radiation, elevated temperatures and drought. Recent predictions of future climate conditions foresee increases in temperature and a decline in rainfall periods at crucial times for cocoa production in the current production zones in West Africa (IPCC, 2021a; Schroth et al., 2016; Stocker et al., 2013). The global average air temperature is expected to increase by between 0.8 and 5.4 °C, while annual rainfall may decline by 1.1–20.5% between 2020 and 2080 depending on the emission scenario (IPCC, 2021b; NCCAS, 2012; Pielke et al., 2022). This is a cause for concern, since elevated temperatures, reduced rainfall, longer dry seasons and higher incidences of pests and diseases are expected to reduce cocoa yields (Cilas & Bastide, 2020; Gachene et al., 2014; Muller et al., 2014, see also Chapter 1).

Breeding resilient varieties has been considered to be a way to increase cocoa yields, especially under future climate scenarios (Vaast & Somarriba, 2014). However, breeding new varieties, especially varieties with increased tolerance to drought and heat, high water-use efficiency and high yields, has been limited by insufficient use of proven breeding methods, limited information on the ecophysiology of cocoa, the plant's long selection cycle, and the heterozygous nature of hybrid parental clones (Efron et al.,

2003). Efforts so far have resulted in hybrid cocoa varieties with increased resistance to pests and diseases and reduced time to maturity (Dos Santos et al., 2014; Frimpong-Anin et al., 2015), but more work is urgently needed on their drought and heat tolerances (Judy et al., 2021). Although marker-assisted selection is being used to study drought-resistant cultivars and genes involved in drought tolerance (Bae et al., 2008), the production and dissemination of cocoa materials that are highly tolerant to drought and heat are still some way off. Selecting drought-tolerant cocoa rootstocks, followed by grafting, is another potential pathway (Zasari et al., 2020).

The provision of shade and the promotion of good agroforestry practices are recommended by many plant scientists to ensure the environmental sustainability of cocoa production (Asare et al., 2017; Asitoakor et al., 2022, Vaast et al., 2016). Agroforestry increases species diversity, provides year-round soil cover and ensures high levels of stored carbon in the soil and in vegetation (Lobĩo et al., 2007; Somarriba et al., 2018). It has also been shown that tree growth and cocoa yields, i.e. the mature productive phase, extend over a longer time span under shade than under full-sun conditions (Ahenkorah et al., 1974). Other benefits of agroforestry include reduced evapotranspiration, enhanced soil fertility and protecting cocoa plants from strong winds and other unfavourable ecological factors (Kyereh, 2017; Miyaji et al., 1997). Furthermore, rates of photosynthesis, growth and yields of cocoa are enhanced under shade (Asare et al., 2018; De Almeida & Valle, 2007; Mensah, 2021). For adult cocoa plants, high yields were observed at shade levels between 30 and 40% (Asare et al., 2018), while about 60% shade is recommended for cocoa seedlings.

In agroforestry systems, companion shade trees in cocoa crop systems have been documented to buffer temperature changes, but they may also have other positive or negative consequences. This depends on the associated tree species that are involved, and whether they lead to root-zone complementarity or competition (Abdulai et al., 2017; Critchley et al., 2022; Jaimes-Suarez et al., 2022; Rigal et al., 2022). Studies of cocoa ecophysiology are difficult because of the size and longevity of cocoa trees, making manipulations difficult. This chapter discusses results from the literature in combination with findings from our on-field studies regarding the effects of shade on cocoa performance under drought and high-temperature stress (Fig. 2.1).

Fig. 2.1 Different cocoa farm configurations and stress trials. **A** Cocoa farm without shade trees. **B** Cocoa agroforestry with remaining shade trees from clearing of the land. **C** Cocoa agroforestry with planted shade trees (*Terminalia sp.* and *Triplochiton scleroxylon*). **D and E** Experiment with a mature cocoa stand under 40% shade using an artificial shade net and with rainwater exclusion. **F** Experiment with cocoa seedlings exposed to heat from non-glowing heaters, with shade (Photos by Eric Opoku Mensah)

The chapter thus draws heavily on two eco-physiological experiments that were conducted in Ghana to study how shade could reduce the effects of drought and elevated temperatures on cocoa physiology (Mensah, 2021). The first experiment took place in the semi-moist region of Ghana investigating the effects of shade and water exclusion on the performance of productive cocoa trees (Fig. 2.1D, E). Plants were monitored over two years for their physiology, growth, litter production and yields. Results from the experiment indicated that shade enhances yields and the physiological performance of cocoa but has limited impacts on water use. The second experiment was conducted at the University of Ghana's Crop Research Farm to test whether shade could reduce the effects of heat on cocoa plants (Mensah et al., 2022). Here, six-month-old cocoa seedlings were exposed to heat provided by 2000W non-glowing infra-red heaters (Fig. 2.1F). The heaters increased the temperature 5–7 °C above the ambient, while 60% shade was provided using black shade nets. Results from the second experiment showed limited effects of shade on the cocoa seedlings under elevated temperature. However, plants kept under shade generally showed enhanced physiology, such as increased chlorophyll fluorescence, chlorophyll pigmentation, stomatal conductance and growth, compared to plants in full-sun conditions.

2.2   Drought and Cocoa Production

Drought is a period in which moisture content in the soil is limited so that plants cannot extract sufficient water for growth and physiological activities (Coder, 1999). It occurs under conditions of low soil and atmospheric humidity when the transpiration flux exceeds the plant uptake of water from the soil. Drought has severe effects on cocoa physiology and restricts stomatal conductance and photosynthesis, and hence vegetative and reproductive plant growth.

2.2.1   Soil Moisture

Soil water content (SWC) is the amount of water present in the soil (Datta et al., 2018). At low SWC, leaves start drooping and may reach the Permanent Wilting Point (PWP), the threshold where plants can no longer recover even if re-watered (Datta et al., 2018).

Cocoa plants have shallow rooting systems (Carr & Lockwood, 2011), with most of the roots concentrated within the first 80 cm of the soil profile, and with over 80% of the root biomass within the top 40 cm, restricting the possibility for water extraction from deep soil layers (Lahive et al., 2019; Moser et al., 2010). The amount of soil water obviously depends on rainfall patterns and evapotranspiration, but also on the soil type and soil depth. For example, clayey soils hold larger amounts of water than sandy soils, and deep soils conserve more available water than shallow soils. In most cocoa-producing countries in West Africa, soil water is depleted in the top 60 cm of soil depth during extended dry seasons, thus exposing the plants to drought (Abdulai et al., 2017).

Under shade, the temperature may fall to 5 °C lower than outside the canopy during the day, maintaining shaded cocoa plants under conditions of relatively high humidity. This means a lower vapour pressure deficit (which is the driving force for transpiration), and it has been suggested that agroforestry reduces cocoa evapotranspiration and allows cocoa to survive under sub-optimal climate conditions (Acheampong et al., 2013; Neither et al., 2018). However, this depends on complementarity in water use between shade tree species and cocoa and hence works best with deep-rooted shade trees that tap soil water below the cocoa root zone. Species selection for cocoa production is very important under drought conditions, as some shade trees, such as *Albizia ferruginea* and *Antiaris toxicaria* (leguminous tree species), have been found to compete with cocoa plants for soil moisture during the dry season (Abdulai et al., 2017; Adams et al., 2016).

2.2.2   *Effects of Drought on Plant-Water Potential*

Water potential is an expression of the water status of a plant, with negative values indicating a relative absence of water. When soil moisture is reduced, roots may not keep up with the pace of evaporation (also known as transpiration) from the leaves, increasing tension in the water-transporting tissues (the xylem) and making plant-water potential more negative. Under conditions of severe drought, the water potential becomes increasingly negative and may cause the formation of air bubbles in the xylem (known as cavitation), which blocks water transport and may in severe cases be lethal to the plant. It is noted that in cocoa, the stem xylem has a larger diameter than the root xylem, which may contribute to plant sensitivity to cavitation under drought (Kotowska et al., 2015).

The plant-water potential affects many physiological processes, and most importantly, it controls the opening and closing of stomata in leaves. Stomata are the pores through which the plant takes up CO_2 and loses water vapour. Under normal, well-watered conditions, cocoa plants will have a water potential ranging between 0.0 and −0.4 MPa (Deloire & Heyns, 2011; Zanetti et al., 2016), whereas values below − 0.8 MPa indicate a water deficit (Deloire & Heyns, 2011). In a through-fall displacement study in Indonesia, after six months of drought, roots experienced declining water potential, falling below −1.5 MPa (Moser et al., 2010) that caused permanent closure of stomata. In a study from Ghana during the dry season, most of the cocoa plants died in response to very low soil moisture because of competition with shade trees (Abdulai et al., 2017).

Shade increases relative humidity around the cocoa plants, thereby reducing transpiration and thus potentially maintaining plant-water potential at a high level. In our field experiment, water exclusion reduced the predawn water potential of cocoa plants, with lower values observed during the dry season (Table 2.1). However, shade resulted in slightly higher water potentials, confirming that shade has a positive impact on the water status of cocoa trees. Reduced plant-water potential in full sun may be the result of increased evapotranspiration resulting from higher leaf temperatures and dryer air.

Table 2.1 Average predawn leaf water potential of cocoa plants at three different levels of rainwater exclusion and two levels of shade measured over two years

Shade	Rainwater exclusion	Predawn water potential (MPa)	
		Rainy season	Dry season
Shade	Full rainwater (0/3W)	−0.24 ± 0.12[a]	−0.40 ± 0.17[a]
	Moderate rainwater exclusion (1/3W)	−0.30 ± 0.13[b]	−0.46 ± 0.15[bc]
	Severe rainwater exclusion (2/3W)	−0.36 ± 0.15[c]	−0.51 ± 0.14[bc]
Sun	Full rainwater (0/3W)	−0.31 ± 0.16[bc]	−0.46 ± 0.16[b]
	Moderate rainwater exclusion (1/3W)	−0.36 ± 0.15[c]	−0.52 ± 0.14[c]
	Severe rainwater exclusion (2/3W)	−0.43 ± 0.17[d]	−0.59 ± 0.14[d]

Note Numbers indicate means ± s.d. ($n = 3$). Means followed by different letters are significantly different according to Tukey's multiple range test ($P < 0.05$)

2.2.3   *Effects of Drought on Photosynthesis*

Photosynthesis, the process by which plants use sunlight, water and carbon dioxide to create oxygen and energy in the form of carbohydrates, is impaired when soil–water content is decreasing (Carr & Lockwood, 2011; Datta et al., 2018; Kirschbaum, 2004). Reduced rates of photosynthesis may be due to partial closure of stomata but can also be due to biochemical limitations (Liang et al., 2019) (see Sect. 2.4). Stomata regulates both transpirational water loss and CO_2 diffusion into the leaves (Barbour, 2016). As discussed above, under drought stress, many plants reduce their stomatal opening to conserve water, at the cost of reducing plant absorption of CO_2 for photosynthesis. The closure of the stomata reduces cooling of the leaves through evaporation, thus increasing leaf temperature. Very high leaf temperatures may harm the leaves and cause leaf wilting. Cocoa plants have low stomatal conductance under water stress and low relative humidity (De Almeida & Valle, 2007) compared to large stomatal opening under non-limiting water conditions and high relative humidity (Sena et al., 1987). Stomatal opening is often assessed in terms of stomatal conductance, a standardized measure of opening. In our field study, stomatal conductance showed a strong seasonal trend, being especially low during the dry season (Fig. 2.2). Surprisingly, the effects of shade vs. sun appeared to have a larger effect on stomatal conductance compared to water exclusion, with shaded cocoa plants having larger stomatal conductance than sun plants.

Despite the differences in stomatal conductance, rates of photosynthesis were comparable between sun and shade plants, with a tendency towards slightly higher values for the former (Fig. 2.2). Since photosynthesis is driven by light, it would be natural to expect a large decrease in photosynthesis in shaded plants. However, in addition to having larger stomatal opening, shaded cocoa plants were able to use the available light and achieve relatively high rates of photosynthesis. Cocoa plants have low light-saturation points, meaning that they reach saturation for photosynthesis at relatively low levels of light, corresponding to 500 μmol photons m^{-2} s^{-1} or ca. 20% of the natural sunlight (Anim-Kwapong & Frimpong, 2004; Salazar et al., 2018). Hence, plants in full sun may not be able to take advantage of the extra radiation available to them. The ability to capture light in shade may also be a result of a reorganization of the photosynthetic system development of large leaves with longer lifespans and increased chlorophyll pigments in the leaves.

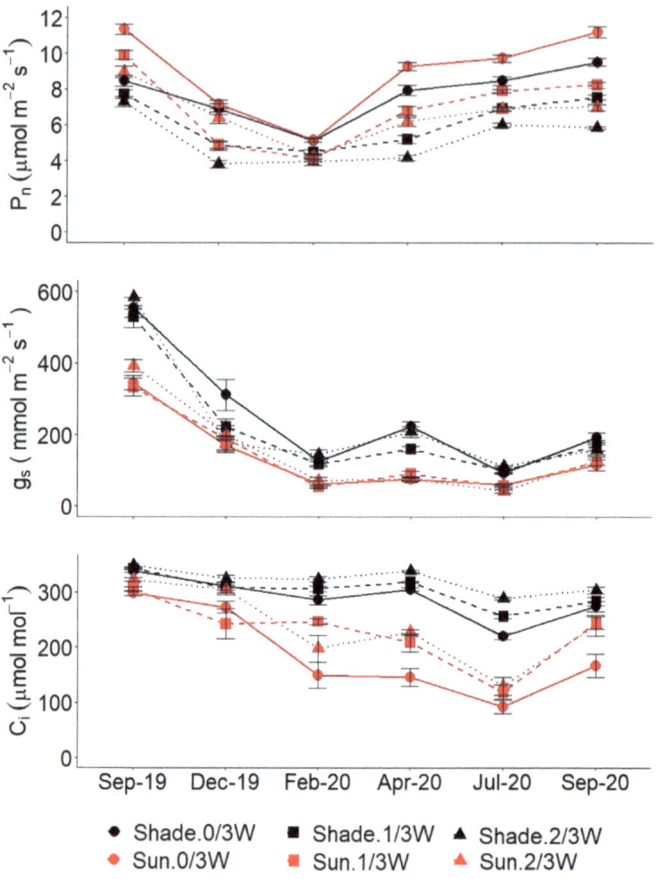

Fig. 2.2 Physiological reactions to shade and drought. Effects of shade and rainwater exclusion on photosynthesis rate (P_n), stomatal conductance (g_s) and sub-stomatal CO_2 concentration (C_i) of a 12-year-old cocoa plant. Codes indicate water availability: 0/3W—full rainwater; 1/3W—partial water exclusion; 2/3W—severe water exclusion

Responses to drought may be dependent on genotypes. Some reports indicate different responses of stomatal conductance and transpiration among cocoa cultivars (Daymond et al., 2011; De Almeida et al., 2015), suggesting that it may be possible to identify cultivars that perform better under drought stress than others. Further research on the varietal differences of stomata regulation and water use in cocoa plants is needed. This also includes studies of whether cocoa has a predominantly anisohydric behaviour (i.e. a variable water content because of continued transpiration at low soil moisture, due to limited stomatal adjustment) or an isohydric tendency (with more stable water contents due to closure of stomata after sensing low soil–water potential).

2.2.4   Biochemical Limitations to Photosynthesis

In addition to limitations caused by light availability and stomatal limitations to the diffusion of CO_2, photosynthesis may also be limited by biochemical factors. The presence of such biochemical limitations can be detected by increased levels of CO_2 inside the leaf (C_i). In our rainwater-exclusion experiment, C_i increased in highly stressed plants compared to non-stressed control plants, and the concentration was proportional to the level of stress (Fig. 2.2). Paradoxically, biochemical limitations may in the first instance be caused by high light and limited diffusion of CO_2, caused by closed stomata (Tholen et al., 2012; Haworth et al., 2018). Energy from high light may be directed to toxic oxygen compounds that will react with enzymes and other substances in the cell, thus reducing the capacity of the plant for photosynthesis. Conversely, high sub-stomatal CO_2 concentrations observed in shade, rather than indicating damage to the photosynthetic system, may be caused by plants maintaining high stomatal conductance and in effect facilitating carbon absorption. Shade thus has positive effects on CO_2 absorption and distribution in the leaves, reflecting increased carboxylation.

The study of sub-stomatal CO_2 concentration is also important because CO_2 gradients within the leaf affect the efficiency of the enzyme fixing CO_2 into sugars (RubisCO) and the nitrogen use efficiency (Evans & von Caemmerer, 1996). Limited information is available on the effect of environmental conditions on sub-stomatal CO_2 concentration in cocoa.

2.3   HEAT AND COCOA

High temperature is one of the main limiting factors for cocoa production (De Almeida & Valle, 2007). High temperature affects the physiology of plants, including the effects of changed stomatal frequency, chlorophyll synthesis (the green pigments in the leaf), enzyme activity and sugar transport (Lamaoui et al., 2018; Wiser et al., 2004).

2.3.1   Photosynthesis

As mentioned, stomata control CO_2 and water movement in and out of the plant through the pore area, the density on the leaf surface and the degree of opening. In cocoa, stomatal densities are higher for leaves developed under mild water stress (Carr & Lockwood, 2011; Huan et al., 1986), but are also higher in leaves developed in full sun compared to shaded leaves (De Almeida & Valle, 2007). In our heat experiment, seedlings in full sun had denser stomata per unit area than seedlings in shade, and heat increased the number of stomata produced per unit area under both full sun and shade. Such differences naturally affect photosynthetic performance, although knowledge on pore size is also needed to accurately assess potential rates of gas flux in and out of leaves.

Most enzymes, including those involved in photosynthesis, work faster with increasing temperatures until they reach the maximum level, where they start uncoiling and lose their function (denaturation). For example, temperatures above 40 °C destroyed the light harvesting complexes in the leaves of perennial plants such as fingered citron and reduced assimilation (Chen et al., 2012; Hasanuzzaman et al., 2013). Another temperature-dependent process affecting rates of photosynthesis is photorespiration. The enzyme fixing CO_2 into sugars, RubisCO, occasionally catalyzes a reaction called photorespiration where O_2 takes the place of CO_2. Photorespiration increases with temperature and leads to a declining net photosynthesis at high temperatures. Furthermore, high temperature inactivates the enzyme system, which transforms sugars into starch, resulting in accumulation of sugars, causing a downregulation of the rate of photosynthesis (Franck et al., 2006; Mathur et al., 2014).

In our heat experiment, we showed that photosynthesis of cocoa is affected by the growing temperature (Fig. 2.3). Temperature optima were between 31 and 33 °C (see also Avila-Lovera et al., 2016; Yapp, 1992) but were almost similar across treatments. The optimum temperature range for photosynthesis coincided with the daily average environmental temperature of 29–33 °C in the experimental site. Having optimum temperature for photosynthesis close to the environmental temperature helps plants to thrive and function well in their environment (Slot & Winter, 2017). Above the optimum, the rate of photosynthesis declines due to photorespiration and, at higher temperatures, the denaturation of enzymes.

On the other hand, the actual levels of photosynthesis were affected by both shade and heat treatments. Photosynthetic capacity was higher for plants growing in full sun compared to shaded plants, and heat reduced the photosynthetic capacity considerably at all temperatures (Fig. 2.3). However, our analysis did not show interactions between sun/shade and heat/no-heat treatments, suggesting that shade could not prevent the loss of photosynthetic capacity caused by the heat treatments (Mensah et al., 2022).

Measurement of chlorophyll fluorescence showed that part of the decrease in photosynthesis seen under heat stress was caused by damages at photosystem II, which is the enzyme complex that fixes the energy from light by removing an electron from oxygen. Chlorophyll fluorescence

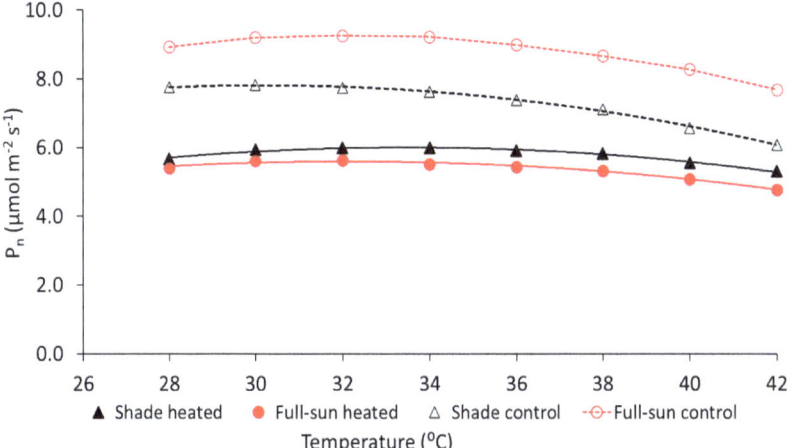

Fig. 2.3 Physiological reactions to high temperature stress. Effects of shade and heat on the photosynthesis rate at different levels of temperature (Adapted from Mensah et al. [2022]. Creative Commons Attribution BY 4.0)

(F_v/F_m) reflects the photochemical activity of photosystem II (PSII) and has previously been used to detect and quantify temperature-induced changes in the photosynthetic system (Chen et al., 2012; Murchie & Lawson, 2013). In our experiment, predawn chlorophyll fluorescence was reduced from 0.80 in control treatments to 0.68 after 28 days of heat imposition, indicating severe stress to the photosynthetic system.

Another cause for lower photosynthesis seems to be a changed concentration and composition of the chlorophylls, the green pigments responsible for capturing light for photosynthesis. We observed reduced leaf chlorophyll contents under heat stress, suggesting impaired chlorophyll biosynthesis (Datta et al., 2009). Also, the ratio between chlorophylls A and B was affected (Mensah et al., 2022). Reduced chlorophyll contents could reduce photosynthesis, resulting in substantial loss in plant productivity. Again, values under shade were higher, but effects were not strong enough to prevent a decrease for the heated seedlings (Salazar et al., 2018).

2.4   Flower and Pod Development Under Heat and Drought Stress

The flowering of cocoa starts eighteen months after planting for some early yielding varieties, while for most varieties, this occurs between three to five years (De Almeida & Valle, 2007). Only 0.5–5% of the flowers develop into mature pods (Carr & Lockwood, 2011). Flowering intensity, pod formation and sizes are affected by drought and heat. Pollen and stigma viability, anthesis, pollen-tube growth and early embryo development are all vulnerable to heat stress (Giorno et al., 2013; Lamaoui et al., 2018). Increased rainfall promotes flushing and flower initiation in cocoa, which is mostly followed by flower and fruit abortion in the dry season (Frimpong-Anin et al., 2014). While flowers and fruits drop during the dry season, mainly because of water stress, flower and fruit abortion in the rainy season is a way for cocoa to manage the resources available for the plants to develop pods (Handley, 2016; Stephenson, 1981). This is affected by plant hormones, the positions of the flowers or the pods on the plant, and rates of cross-pollination (Carr & Lockwood, 2011; Handley, 2016).

Cocoa attains full potential yield between eight to ten years after planting (De Almeida & Valle, 2007). The average yield is between 300 and 500 kg ha^{-1} in West Africa (Bymolt et al., 2018) corresponding to only a third of the potential yield (Aneani & Ofori-Frimpong, 2013; MOFA, 2016). The low yields seem to be partly due to limitations in water supply (Asante et al., 2022). In our study on shade and water-exclusion effects, yields were generally between 200 and 700 kg ha^{-1} per year depending on the suppression and/or the shade levels. Under full rainwater, shade increased yields by about 23% compared to full-sun conditions, while severe water exclusion reduced yields to as low as 59%. While shade was beneficial under all levels of water supply, it was not sufficient to prevent lower yields when water was restricted.

2.5   Conclusion

In West Africa, climate change is already having negative impacts on cocoa production and therefore on cocoa farmers' livelihoods. All stakeholders along the cocoa value chain (from cocoa farmers to purchasing companies) are increasingly being affected, in Ghana as well as in the neighbouring cocoa-producing countries. Field performance and yields are expected to be reduced further due to increasing rainfall variability, longer dry seasons and rising temperatures, making climate change the key challenge faced by cocoa producers. Our results confirm that drought and high temperatures have negative impacts on cocoa physiology, leading to reduced yields. Shade, on the other hand, improves both physiological performance and yields, thus confirming on-field research suggesting that agroforestry systems may increase yields (see Chapter 3). Although we recommend cultivation under shade, shade alone does not reduce the negative impact of stresses sufficiently to prevent damage from extreme climate change. Hence, while agroforestry represents an overall benefit under medium to high rainfall conditions, it will be necessary to refine agroforestry management using other climate-smart management innovations to improve the performance of cocoa under the expected climate change. This could include the following:

- Exploring the effects of irrigation systems under shade
- Selecting and managing tree species according to the local context
- Selecting cocoa rootstocks and varieties that are highly performant under water-limiting conditions
- Studying shade tree-cacao interactions to understand cocoa physiology under agroforestry conditions

- Identifying deep-rooted shade trees with limited competition with cocoa plants' root zones under drought stress.

Aside from climate, yields are influenced by a wide range of factors, including labour costs, diseases and pests, soil fertility, choice of cocoa variety, the age of cocoa trees, and the age and training of farmers (Abdulai et al., 2020). Higher yields may be possible with the use of technologies such as fertilization, pest and disease control, timely harvesting, pruning, supplemental irrigation and planting high-yielding cocoa cultivars (Laven & Boomsma, 2012). Any intervention regarding the use of shade will have to consider these factors, which are explored in more detail in other chapters of this book.

References

Abdulai, I., Hoffmann, M. P., Jassogne, L., Asare, R., Graefe, S., Tao, H.-H., Muilerman, S., Vaast, P., Asten, P. V., & Laderach, P. (2020). Variations in yield gaps of smallholder cocoa systems and the main determining factors along a climate gradient in Ghana. *Agricultural Systems, 181*, 102–812.

Abdulai, I., Vaast, P., Hoffman, M., Asare, R., Jassogne, L., Asten, V. P., Rotter, P. R., & Graefe, S. (2017). Cocoa agroforestry is less resilient to sub-optimal and extreme climate than cocoa in full sun. *Global Change Biology, 24*(1), 273–286.

Acheampong, K., Hadley, P., & Daymond, A. J. (2013). Photosynthetic activity and early growth of four cacao genotypes as influenced by different shade regimes under West Africa dry and wet season conditions. *Experimental Agriculture, 49*(1), 31–42.

Adams, M. A., Turnbulla, T. L., Sprent, J. I., & Buchmannc, N. (2016). Legumes are different: Leaf nitrogen, photosynthesis, and water use efficiency. *PNAS, 113*, 4098–4113.

Ahenkorah, Y., Akrofi, G. S., & Adri, A. K. (1974). The end of the first cocoa shade and manurial experiment at the Cocoa Research Institute of Ghana. *Journal of Horticultural Science, 49*, 43–51.

Ameyaw, L. K., Ettl, G. J., Leissle, K., & Anim-Kwapong, G. J. (2018). Cocoa and climate change: Insights from smallholder cocoa producers in Ghana regarding challenges in implementing climate change mitigation strategies. *Forest, 9*(742), 1–20.

Anim-Kwapong, G. J., & Frimpong, E. B. (2004). *Vulnerability and adaptation assessment under The Netherlands Climate Change Studies Assistance Programme Phase 2 (NCCSAP2): Vulnerability of agriculture to climate change-impact of climate on cocoa production* (Vol. 2). Cocoa Research Institute of Ghana.

Aneani, F., & Ofori-Frimpong, K. (2013). An analysis of yield gap and some factors of cocoa (Theobroma cacao) yields in Ghana. *Sustainable Agricultural Research, 2*(4), 117–127.

Asante, P. A., Rahn, E., Zuidema, P. A., Rozendaal, M. A., van der Baan, M. E. G., Laderah, P., Asare, R., Cryer, C. N., & Anten, N. P. R. (2022). The cocoa yield gap in Ghana: A quantification and an analysis of factors that could narrow the gap. *Agricultural Systems, 201*, 103473.

Asare, R., Asare, R., Asante, W., Markussen, B., & Raebild, A. (2017). Influences of shading and fertilization on on-farm yields of cocoa in Ghana. *Experimental Agriculture, 53*(3), 416–431.

Asare, R., Bo, M., Asare, A. R., Anim-Kwapong, G., & Rabild, A. (2018). On-farm cocoa yields increase with canopy cover of shade trees in two agro-ecological zones in Ghana. *Climate and Development, 11*(5), 435–445.

Asitoakor, B. K., Vaast P., Rabild, A., Ravn, H. P., Eziah, V. Y., Owusu, K., Mensah, E. O., & Asare, R. (2022). Selected shade tree species improved cocoa yields in low-input agroforestry systems in Ghana. *Agricultural Systems, 202*, 103476.

Avila-Lovera, E., Cornel, I., Jaimez, R., Urich, R., Pereyra, G., Araques, O., Chacon, I., & Tezara, W. (2016). Ecophysiological traits of adult trees of Criollo cocoa cultivars (*Theobroma cacao L.*) from a germplasm bank in Venezuela. *Experimental Agriculture, 52*(1), 137–153.

Bae, H., Kim, S.-H., Kim, M. S., Sicher, R. C., Lary, D., Strem, M. D., Natarajan, S., & Bailey, A. B. (2008). The drought response of *Theobroma cacao* (cacao) and the regulation of genes involved in polyamine biosynthesis by drought and other stresses. *Plant Physiology and Biochemistry, 46*, 174–188.

Barbour, M. M. (2016). Understanding regulation of leaf internal carbon and water transport using online stable isotopes techniques. *New Phytologist, 213*, 83–88.

Bartley, B. G. D. (2005). *The genetic diversity of Cacao and its utilization.* CABI Publishing.

Bymolt, R., Laven, A., & Tyszler, M. (2018). Production, and yield. In A. Laven, R. Bymolt, & M. Tyszler (Eds.), *Demystifying the cocoa sector in Ghana and Côte d'Ivoire* (pp. 194–206). The Royal Tropical Institute (KIT).

Carr, M. K. V., & Lockwood, G. (2011). The water relations and irrigation requirements of cocoa (Theobroma cacao L.): A review. *Experimental Agriculture, 47*(4), 653–676.

Cheesman, E. E. (1944). Notes on the nomenclature, classification, and possible relationships of cacao populations. *Tropical Agriculture, 21*, 144–159.

Chen, W. R., Zheng, J. S., Li, Y. Q., & Guo, W. D. (2012). Effects of high temperature on photosynthesis, chlorophyll fluorescence, chloroplast ultra-structure and antioxidant activities in fingered citron. *Russian Journal of Plant Physiology, 59*(6), 732–740.

Cilas, C., & Bastide, P. (2020). Challenges to cocoa production in the face of climate change and the spread of pests and diseases. *Agronomy, 10*(1232), 1–8.

Coder, K. D. (1999). *Drought damage to trees*. Daniel B. Warnell School of Forest Resources, University of Georgia (Extension publication).

Critchley, M., Sassen, M., Rahn, E., Ashiagbior, G., Soesbergen, A., & Maney, C. (2022). *Identifying opportunity areas for cocoa agroforestry in Ghana to meet policy objectives*. United Nations Environmental Programme of World Conservation Monitoring Centre.

Datta, S., Mohanty, S., & Tripathy, C. (2009). Role of temperature stress on chloroplast biogenesis and protein import in pea. *Plant Physiology, 150*, 1050–1061.

Datta, S., Stivers, J., & Taghvaeian, S. (2018). *Understanding soil water content and thresholds for irrigation management*. OSU Extension Fact Sheets.

Daymond, A. J., Tricker, P. J., & Hadley, P. (2011). Genotypic variation in photosynthesis in cacao is correlated with stomatal conductance and leaf nitrogen. *Biologia Plantarum, 55*(1), 99–104.

De Almeida, A.-A. F., & Valle, R. R. (2007). Ecophysiology of the cacao tree. *Brazilian Journal of Plant Physiology, 19*(4), 425–448.

De Almeida, J., Wilmer, T., & Herrara, A. (2015). Physiological response to drought and experimental water deficit and waterlogging of four clones of cacao (*Theobroma cacao* L.) selected for cultivation Venezuela. *Agricultural Water Management, 171*, 80–88.

Deloire, A., & Heyns, D. (2011). *The leaf water potentials: Principles, methods, and thresholds* (pp 129–131). Vineyard. Retrieved on 12 September 2017, from https://www.researchgate.net/publication/259589 941_The_leaf_water_potentials_principles_method_and_thresholds

Dos Santos, I. C., de Almeida, A.-A. F., Anhert, D., da Canceicao, A. S., Pirovani, C. P., Pires, L. J., Valle, R. R., & Baligar, V. C. (2014). Molecular, physiological, and biochemical responses of Theobroma cacao L. genotypes to soil water deficit. *PloS ONE, 9*(12), Article e115746.

Efron, Y., Epaina, P., & Marfu, J. (2003). *Breeding strategies to improve cocoa production in Papua New Guinea*. International Workshop on Cocoa Breeding for Improved Production Systems, pp. 12–32.

Evans, J. R., & von Caemmerer, S. (1996). Carbon dioxide diffusion inside leaves. *Plant Physiologist, 110*, 339–346.

Franck, N., Vaast, P., Génard, M., & Dauzat, J. (2006). Soluble sugars mediate sink feedback down-regulation of leaf photosynthesis in field-grown *Coffea arabica*. *Tree Physiology, 26*(4), 517–525.

Frimpong-Anin, K., Adjaloo, K. M., Kwapong, P. K., & Oduro, W. (2014). Structure and stability of cocoa flowers and their response to pollination. *Journal of Botany, 2014*, 1–6.

Frimpong-Anin, K., Bosu, P. P., Adjaloo, K. M., Braimah, H., Oduro, W., & Kwapong P. K. (2015). *Some facts about cocoa pollination*. University of Cape Coast Printing Press.

Gachene, K. K. C., Karuma, A. N., & Baaru, M. W. (2014). Climate change and crop yield in Sub-Saharan Africa. In R. Lal, B. Singh, D. Nwaseba, D. Kraybill, D. Hansen, & L. Eik (Eds.), *Sustainable intensification to advance food security and enhance climate resilience in Africa* (pp. 165–183). Springer.

Giorno, F., Wolters-Arts, M., Mariani, C., & Rieu, I. (2013). Ensuring reproduction at high temperatures: The heat stress response during anther and pollen development. *Plants, 2*, 489–506.

Handley, L. R. (2016). The effects of climate change on the reproductive development of *Theobroma cacao L.* (PhD dissertation). University of Readings, Readings.

Haworth, M., Marino, G., Brunetti, C., Killi, D., De Carlo, A., & Centritto, M. (2018). The impact of heat stress and water deficit on the photosynthetic and stomatal physiology of olive (*Olea eurpaea* L.): A case study of the 2017 heat wave. *Plants, 7*(76), 1–13.

Hasanuzzaman, M., Nahar, K., & Fujita, M. (2013). Extreme temperature responses, oxidative stress, and antioxidant defense in plants. In K. Vahdati & C. Leslie. (Eds.), *Abiotic stress—Plant responses and applications in agriculture*. Intech. https://doi.org/10.5772/54833

Huan, L. K., Yee, H. C., & Wood, B. J. (1986). Irrigation of cocoa on coastal soils in Peninsular Malaysia. *Cocoa and coconuts: Progress and outlook* (pp. 117–132). Incorporated Society of Planters.

ICCO (International Cocoa Organization). (2020). *Growing cocoa*. Retrieved on 20 October 2022, from https://www.icco.org/growing-cocoa/

IPCC (Intergovernmental Panel on Climate Change). (2021a). *Climate change 2021: The physical science basis. Contribution of working group I to the sixth assessment report of the Intergovernmental Panel on Climate Change (IPCC)*.

IPCC (Intergovernmental Panel on Climate Change). (2021b). Summary of policymakers. In: *Climate change 2021: The physical science basis. Contribution of working group 1 to the sixth assessment report of the Intergovernmental Panel on Climate Change* (IPCC AR6 WGI). Cambridge University Press.

Jaimes-Suarez, Y. Y., Carvajal-Rivera, A. S., Galvis-Neira, D. A., Carvalho, F. E. L., & Rojas-Molina, J. (2022). Cacao agroforestry systems beyond the stigmas: Biotic and abiotic stress incidence impact. *Frontiers in Plant Science, 13*, 921469.

Judy, B., Mimimol, J. S., Suma, B., Santhoshkumar, A. V., Jiji, J., & Panchami, P. S. (2021). Drought mitigation in cocoa (*Theobroma cacao* L.) through developing tolerant hybrids. *BMC Plant Biology, 21*(594), 1–12.

Kirschbaum, M. U. F. (2004). Direct and indirect climate change effects on photosynthesis and transpiration. *Plant Biology, 6*(3), 242–253.

Kotowska, M. M., Hertel, D., Rajab, Y. A., Barus, H., & Schuldt, B. (2015). Patterns in hydraulic architecture from roots to branches in six tropical tree species from cacao agroforestry and their relation to wood density and stem growth. *Frontiers of Plant Science, 6*(191), 1–16.

Kyereh, D. (2017). Shade trees in cocoa agroforestry systems in Ghana: Influence on water and light availability in dry seasons. *Journal of Agriculture and Ecology Research International, 10*(2), 1–7.

Lahive, F., Hadley, P., & Daymond, A. J. (2019). The physiological response of cacao to the environment and the implications for climate change resilience. A review. *Agronomy of Sustainable Development, 39*(5), 1–22.

Lamaoui, M., Jemo, M., Datla, R., & Bekkaoui, F. (2018). Heat and drought stresses in crops and approaches for their mitigation. *Frontiers in Chemistry, 6*(26), 1–12.

Laven, A., & Boomsma, M. (2012). *Incentives for sustainable cocoa production in Ghana. Moving from maximizing outputs to optimizing performance*. Royal Tropical Institute.

Liang, G., Liu, J., Zhang, J., & Guo, J. (2019). Effects of drought stress on photosynthetic and physiological parameters of tomato. *America Society for Horticultural Sciences, 145*(1), 12–17.

Lobĩo, D. E., Setenta, W. C., Lobĩo, E. S. P, Curvelo, K., & Valle R. R. (2007). Cacao Cabrucal, sistema agrossilvicultural tropical. In R. R. Valle (Eds.), *Ciencia, Tecnologia e Manejo do Cacaueiro* (pp. 290–323). Grafica e Editora Vital Ltda.

Loor, R. G., Risterucci, A. M., Courtois, B., Fouet, O., Jeanneau, M., Rosenquist, E., Amores, F., Vasco, A., Medina, M., & Lanaud, C. (2009). Tracing the native ancestors of the modern Theobroma cacao L. population in Ecuador. *Tree Genetics and Genomes, 5*, 421–433.

Mattayasovszky, M. (2017). *Top 10 cocoa producing countries*. WorldAtlas, Reunion Technology Inc

Mathur, S., Agrawal, D., & Jajoo, A. (2014). Photosynthesis: Response to high temperature stress. *Journal of Photochemistry and Photobiology b: Biology, 137*, 116–126.

Miyaji, K. I., da Silva, W. S., & Alvim, P. de T. (1997). Longevity of leaves of a tropical tree: Theobroma cacao grown under shading in relation to emergence. *New Phytologist, 135*, 445–454.

Mensah, E. O., Asare, R., Vaast, P., Amoatey, C. A., Markussen, B., Owusu, K., Asitoakor, B. K., & Rabild, A. (2022). Limited effects of heat and shade on cocoa (*Theobroma cacao* L.) physiology. *Environmental and Experimental Botany, 201*, 1–11.

Mensah, E. O. (2021). *Effect of shade on ecophysiology of cocoa under stress conditions* (PhD thesis). University of Ghana, Accra, Ghana.

MOFA (Ministry of Food and Agriculture). (2016). *Agriculture in Ghana: Facts and figures*. Statistics, Research, and Information Directorate of MOFA.

Moser, G., Leuschner, C., Hertel, D., Hoilscher, D., Kohler, M., Leitner, D., Michalzik, B., Prihastanti, E., Tjitrosemito, S., & Schwendenmann, L. (2010). Response of cocoa trees (*Theobroma cacao*) to a 13-month desiccation period in Sulawesi, Indonesia. *Agroforest System, 79*, 171–187.

Muller, C., Waha, K., Bondeau, A., & Heinke, J. (2014). Hotspots of climate change impacts in sub-Saharan Africa and implications for adaptation and development. *Global Change Biology, 20*, 2505–2517.

Murchie, E. H., & Lawson, T. (2013). Chlorophyll fluorescence analysis: A guide to good practice and understanding some new applications. *Journal of Experimental Botany, 64*(13), 3983–3998.

NCCAS (Ghana National Climate Change Adaptation Strategy). (2012). *National Climate change adaptation strategy*. Retrieved on 27 October 2017, from http://adaptation-undp.org/sites/default/files/downloads/ghana_national_climate_change_adaptation_strategy_nccas.pdf

Neither, W., Armengot, L., Andres, C., & Schneider, M. (2018). Shade trees and tree pruning alter throughfall and microclimate in cocoa (Theobroma cacao L.) production systems. *Annals of Forest Sciences, 75*(38), 1–16.

Pielke, R., Jr., Buress, M. G., & Ritchie, J. (2022). Plausible 2005–2050 emissions scenarios project between 2 °C and 3 °C of warming by 2100. *Environmental Research Letters, 17*(2), 1–8.

Rigal, C., Wagner, S., Nguyen, M. P., Jassogne, L., & Vaast, P. (2022). Shade tree advice methodology: Guiding tree species selection using local knowledge. *People and Nature, 4*, 1233–1248.

Salazar, S. C. J., Melgarejo, L. M., Casanoves, F., Rienzo, J. D. A. and Damatta, C. A. (2018). Photosynthesis limitations in cacao leaves under different agroforestry systems in the Colombian Amazon. *PLoS ONE, 13*(11), Article e0206149.

Schroth, G., Läderach, P., Martinez-Valle, A. I., Bunn, C., & Jassogne, L. (2016). Vulnerability to climate change of cocoa in West Africa: Patterns, opportunities, and limits to adaptation. *Science of the Total Environment, 556*, 231–241.

Sena, G. A. R., Kozlowski, T. T., & Reich, P. B. (1987). Some physiological responses of *Theobroma cacao* var. catongo seedlings to air humidity. *New Phytologist, 107*, 591–602.

Slot, M., & Winter, K. (2017). In situ temperature response of photosynthesis of 42 tree and liana species in the canopy of two Panamanian lowland tropical forests with contrasting rainfall regimes. *New Phytologist, 214*, 1103–1117.

Somarriba, J. E., Orozco-Agullar, L., Cerda, R. C., & Sampson, A. L. (2018). *Analysis and design of the shade canopy of cocoa-based agroforestry systems.* Burleigh Dodds Science Publishing Limited.

Stephenson, A. (1981). Flower and fruit abortion: Proximate causes and ultimate functions. *Annual Review of Ecology and Systematics, 12*, 253–279.

Stocker, T, F., Qin, D., Pattner, G.-K., et al. (2013). *Climate change 2013: The physical science basis working group 1 contribution to the fifth assessment of the Intergovernmental Panel on Climate Change—Abstract for decision-makers* (pp. 19–23). Intergovernmental Panel on Climate Change.

Tee, Y. K., Abdul Haadi, H., Raja, N. A., & Mohd, R. S. (2018). Stress tolerance of cacao trees (Theobroma cacao L.) subjected to smart water gel. Malaysian Society of Plant Physiology Conference Trans. Malaysian Soc. *Plant Physiology, 25*, 1–6

Tholen, D., Ethier, G., Genty, B., Pepin, S., & Zhu, X.-G. (2012). Variable mesophyll conductance revisited: Theoretical background and experimental implications. *Plant, Cell and Environment, 35*, 20187–22103.

Umaharan, P. (2018). *Achieving sustainable cultivation of cocoa.* Cocoa Research Centre, The University of West Indies.

Vaast, P., Harmand J. M., Rapidel B., Jagoret P., & Deheuvels O. (2016). Coffee and cocoa production in agroforestry: A climate-smart agriculture model. In T. Emmanuel (Ed.), M. David & C. Paul (Trans.), *Climate change and agriculture worldwide* (pp. 197–208). Springer.

Vaast, P., & Somarriba, E. (2014). Trade-offs between crop intensification and ecosystem services: The role of agroforestry in cocoa cultivation. *Agroforestry Systems, 88*, 947–956.

Wiser, R. R., Olson, A. J., Schrader, S. M., & Sharkey, T. D. (2004). Electron transport is the functional limitation of photosynthesis in field-grown Pima cotton plants at high temperature. *Plant Cell and Environment, 27*(6), 717–724.

Wood, G. A. R., & Lass, R. A. (1992). *Cocoa.* Tropical Agriculture Series (4th ed.). Longman Press.

Yapp, J. H. H. (1992). *A study into the potential for enhancing productivity in cocoa (Theobroma cacao L.) through exploitation of physiological and genetic variation* (Dissertation). University of Reading, Readings, UK.

Zanetti, L. V., Milanez, C. R. D., Gama, V. N., Aguilar, M. A., Souza, C. A. S., Campostrini, E., Ferraz, T. M., & Figueiredo, F. A. M. M. A. (2016). Leaf application of silicon in young cacao plants subjected to water deficit. *Pesquisa Agropecuária Brasileira, 51*(3), 215–223.

Zasari, M., Wachjar, A., Susilo, A. W., & Subarsono, S. (2020). Prope legiti-
mate rootstocks determine the selection criteria for drought-tolerant cocoa.
Biodiversitas, 21(9), 4067–4075.

Open Access This chapter is licensed under the terms of the Creative Commons
Attribution 4.0 International License (http://creativecommons.org/licenses/
by/4.0/), which permits use, sharing, adaptation, distribution and reproduction
in any medium or format, as long as you give appropriate credit to the original
author(s) and the source, provide a link to the Creative Commons license and
indicate if changes were made.

The images or other third party material in this chapter are included in the
chapter's Creative Commons license, unless indicated otherwise in a credit line
to the material. If material is not included in the chapter's Creative Commons
license and your intended use is not permitted by statutory regulation or exceeds
the permitted use, you will need to obtain permission directly from the copyright
holder.

CHAPTER 3

Shade Tree Species Matter: Sustainable Cocoa-Agroforestry Management

*Bismark Kwesi Asitoakor*⑩*, Anders Ræbild*⑩*,*
*Philippe Vaast*⑩*, Hans Peter Ravn*⑩*, Kwadwo Owusu*⑩*,*
*Eric Opoku Mensah*⑩*, and Richard Asare*⑩

Abstract Shade trees are important components of cocoa-agroforestry systems because they influence yields, soil fertility and the occurrence of pests and diseases and may support adaptation to climate change. Based on a review of the existing literature and on primary data from field experiments, this chapter reports on the species-specific effects of shade trees in relation to the management of insect pests, black pod diseases and their impacts on cocoa yield. Shade tree species in cocoa

B. K. Asitoakor (✉) · E. O. Mensah
Department of Crop Science, University of Ghana, Accra, Ghana
e-mail: bkasitoakor001@st.ug.edu.gh; bka@ign.ku.dk

E. O. Mensah
e-mail: omedjin@gmail.com

B. K. Asitoakor · A. Ræbild · H. P. Ravn · E. O. Mensah
Department of Geosciences and Natural Resource Management, University of Copenhagen, Frederiksberg, Denmark
e-mail: are@ign.ku.dk

© The Author(s) 2024
M. F. Olwig et al. (eds.), *Agroforestry as Climate Change Adaptation*,
https://doi.org/10.1007/978-3-031-45635-0_3

systems impact soil available phosphorus differently and shade tree species such as Spanish cedar (*Cedrela odorata*), limba (*Terminalia superba*) and mahogany (*Khaya ivorensis*) increase cocoa yield compared with cocoa systems without shade trees. The architecture of shade tree species may influence below-canopy temperatures and relative humidity, which potentially affect pests such as mirids and black pod disease infections and ultimately cocoa yield. As farmers have local knowledge of and preferences for certain shade tree species, strengthening the combination of scientific and local knowledge can prove a powerful tool for the improved management of shade tree species, as well as cocoa pests and diseases.

Keywords Pests and diseases · Black pod disease · Mirid · Soil fertility · Yield · Climate change

H. P. Ravn
e-mail: hpr@ign.ku.dk

B. K. Asitoakor · E. O. Mensah
CSIR-Plant Genetic Resources Research Institute, Bunso, Ghana

P. Vaast
UMR Eco & Sols, Centre de Coopération Internationale en Recherche Agronomique Pour Le Développement (CIRAD), Université Montpellier, Montpellier, France

World Agroforestry Center, Nairobi, Kenya

P. Vaast
e-mail: philippe.vaast@cirad.fr

K. Owusu
Department of Geography and Resource Development, University of Ghana, Accra, Ghana
e-mail: kowusu@ug.edu.gh

R. Asare
International Institute of Tropical Agriculture (IITA), Accra, Ghana
e-mail: r.asare@cgiar.org

3.1 Introduction

In the tropical regions where cocoa (*Theobroma cacao* L.) is cultivated, many different factors may result in low yield and reduced revenues from the production of the world's raw material for chocolate. These include poor farming practices, the occurrence of pests and diseases and worsening weather conditions due to climate change. Cocoa farmers and cocoa-producing countries must identify strategies that support sustainable production. In addition to improving yield through the development of high-yielding and disease-resistant cocoa varieties (Edwin & Masters, 2005; Mcelroy et al., 2018) and improving fertilizer regimes (Hoffmann et al., 2020; Niether et al., 2019), agroforestry has been recognized as an important means to improve cocoa yield (Asitoakor et al., 2022a). Agroforestry, the deliberate cultivation of crops with forest or food trees, is generally more environmentally friendly than monocropping systems and serves as an important climate change adaptation measure, especially for Sub-Saharan Africa, where most of the global cocoa production takes place (Vaast et al., 2016).

Sustaining cocoa yield is a major challenge for smallholder farmers, especially under worsening climatic conditions with reduced rainfall and increasing temperatures (see Chapter 1). Smallholder farmers lack the capacity to irrigate and afford the required inputs, labour and other agronomic support needed to achieve high yield. Currently, rainfall is erratic in most of the West African cocoa region and below the optimal ranges of 1,500–3,000 mm for cocoa production (Abdulai et al., 2020; IITA, 2009). Temperatures in these areas are increasing (Ruf, 2011; Tscharntke et al., 2011) above the optimal annual maximum of 30–32 °C. Under high temperature and low rainfall conditions, cocoa phenology and performance with regard to flowering and fruiting are impeded (Adjaloo et al., 2012; Asitoakor et al., 2022b; Daymond & Hadley, 2008; Medina & Laliberte, 2017, see also Chapter 2) and insect infestations and the proportions of small size (low-grade) and defective beans increase (Asante-Poku & Angelucci, 2013). In major cocoa areas in Ghana and Côte d'Ivoire, low yield due to pests and diseases, high input demands and the high cost and low availability of labour leave farmers with the question of whether or not to replant their cocoa plots with other crops such as oil palm (*Elaeis guineensis*) and rubber (*Hevea brasiliensis*) (Cocoa Barometer, 2022; Ruf, 2015) or shift to other forms of land use (see Chapter 4). However, the adoption of agroforestry with selected beneficial shade trees might prove cost-effective, preserve the environment and sustain yield (Asare, 2016; Babin et al., 2010; Ofori-Frimpong et al., 2007; Tscharntke et al., 2011; van Vliet et al., 2015, see also Chapter 5).

The scientific debate on the role of agroforestry in cocoa production has been going on for decades, with many arguing that the advantages of shade trees in cocoa systems outweigh their disadvantages, particularly when tree species that are adapted to the local social and agroecological contexts are adequately managed. Benefits from shade trees include carbon sequestration, biodiversity conservation, alternative income for farmers, soil improvement, prevention of erosion and the management of micro-climatic conditions that, among other things, reduce pest and disease infestations (Abdulai et al., 2018; Niether et al., 2019; Tscharntke et al., 2011). However, some tree species serve as alternative hosts for pests (e.g. mirids) and diseases in cocoa systems (Mahob et al., 2015) and they may also compete for nutrients, water and sunlight (van Vliet & Giller, 2017). Common species of shade trees in West Africa include fruit trees such as avocado (*Persea americana*), orange (*Citrus sinensis*), coconut (*Cocos nucifera*) and mango (*Mangifera indica*), as well as timber-producing species such as mahogany (*K. ivorensis*), ceiba (*Ceiba pentandra*) and teak (*Tectona grandis*) (Rigal et al., 2022).

Shade trees are important sources of food and local pharmaceutical raw materials for curing diverse illnesses and diseases (Rao et al., 2004). For example, cola nuts (*Cola nitida*) provide an important ingredient in beverages such as Coca Cola and Pepsi Cola, and are used for short-term relief from fatigue, depression, chronic fatigue syndrome (CFS) and melancholy (Atolani et al., 2019). Flat-crown tree (*Albizia adianthifolia*) is another important shade tree also used for treating diabetes, headaches, eye problems, wounds, pain, skin diseases, gastrointestinal problems, haemorrhoids, infertility in women, respiratory problems and sexually transmitted diseases (Lemmens, 2007a). The bark decoction of mahogany (*K. ivorensis*) is used in the treatment of coughs, fever, malaria, anaemia, wounds, sores, ulcers, tumours, rheumatic pains and lumbago (Lemmens, 2008). Aside from the medicinal role of common shade tree species, the wood from species like stoolwood (*Alstonia boonei*), Spanish cedar (*C. odorata*), African teak (*Milicia excelsa*) and black afara (*Terminalia ivorensis*) are used for construction and furniture, including the building of canoes, roofing and household items such as stools, boxes, tables and chairs (Adotey et al., 2012; Foli, 2009; Lemmens, 2008; Ofori, 2007).

Farmers have clear ideas about the tree species they prefer and the types of shade trees that may provide different types of ecosystem services (Rigal et al., 2022). Farmers' reasons for selecting specific shade tree species include their influence on cocoa yield, their income-generating potential, their medicinal properties, their use in construction and whether they serve as sources of fuelwood (Appendix). Nonetheless, the ecological interactions between shade tree species, pests and diseases and their influence on cocoa yield are poorly known. Few studies have documented the varied impacts of different shade tree species on cocoa production (Abdulai et al., 2018; Asare et al., 2019; Graefe et al., 2017). Asare et al. (2019), in an on-farm study conducted in the Ashanti and Western regions of Ghana to understand the relationship between the canopy cover of shade trees and fertilizer regimes on yield, observed a doubling of yield, as shade cover increased from zero to 30% in 86 plots. The study further showed fertilizer application to have increased yield by 7%. In a study by Kaba et al. (2020), African tulip tree (*Spathodea campanulata*), limba (*T. superba*) and black afara (*T. ivorensis*) were the most desirable tree species, while stoolwood (*A. boonei*) was the least desired in cocoa systems in Ghana's semi-deciduous rainforest zone from the farmers' perspective. The different species' desirability was linked to their influence on cocoa and other food crops around shade trees, the suitability of shade trees as fodder and the general improvement in pod and cocoa-tree health. Graefe et al. (2017) identified *T. ivorensis*, *T. superba*, *M. excelsa*, *A. boonei* and *Pycnanthus angolensis* (African nutmeg) as the five most desired shade tree species by cocoa farmers across Ghana's cocoa belt. The five species were preferred to other species for their compatibility with cocoa, as they were perceived to provide the right amount of shade, improve soil moisture and fertility, have a fast rate of leaf decomposition and suppress weeds. Shade species such as the African corkwood tree (*Musanga cecropioides*), Ceiba (*C. pentandra*), Akee apple tree (*Blighia sapida*), African crabwood (*Carapa procera*) and giant cola (*Cola gigantea*) were assessed by farmers to be less desirable due to their heavy shade, below-ground competition, slow leaf decomposition, being an alternative host for pests and diseases, and causing physical damage to cocoa. Abdulai et al. (2018) observed the common use of gliricidia (*Gliricidia sepium*), avocado (*P. americana*), orange (*C. sinensis*) and the boundary tree (*Newbouldia laevis*) in the mid- and wet cocoa regions in Ghana. According to the authors, G. *sepium* was considered important for soil improvement, *P. americana* and *C. sinensis* for food and *N. laevis* for use as live stakes for yam (*Dioscorea* sp.). In our study, as will be discussed further below, *C. odorata*, *T. superba* and *K. ivorensis* are identified as good shade tree species, associated with more than 40% higher yield of cocoa compared with unshaded cocoa systems (Asitoakor, 2021). Conversely, species like *A. boonei* are viewed differently from place to place, perhaps depending on the specific needs and uses of the species, aside from shade provision.

In many instances, the relationship between shade trees and cocoa yields is attributed to the variations in shade tree structure and growth rates that affect their interactions with cocoa trees (Asante et al., 2021; Asitoakor et al., 2022a). However, there is a need for more knowledge concerning the properties of shade tree species to improve shade tree selection in cocoa-agroforestry systems. This chapter draws on the existing literature and the results of a three-year on-farm experimental study focusing on how eight different agroforestry shade tree species influenced soil nutrients, mirids (*Sahlbergella singularis* Hagl. and *Distantiella theobroma* Dist.) and black pod diseases (caused by *Phythophthora palmivora*, and *P. megakarya*) infections. The next section, Sect. 3.2, highlights the influences of the selected shade tree species on soil fertility and yields in cocoa systems, while Sect. 3.3 shows how the selected shade tree species affected cocoa pests and disease infestations.

3.2   Role of Shade Trees in Soil Fertility and Yield in Cocoa-Agroforestry Systems

Soil fertility may be defined as the capacity of the soil to support the growth and yield of plants (Young, 1990). As with other crops, cocoa yield is directly related to soil fertility, age of the cocoa plant, prevailing climatic conditions (in terms of rainfall, temperature and relative humidity) and agronomic practices (Asante et al., 2021; Asitoakor et al., 2022a). In addition to these factors, the type of shade tree species intercropped on cocoa farms influence the performance of the cocoa plants (Asitoakor et al., 2022a). Though the fertility of cocoa soils varies by location, landscape and number of years under cultivation, the maintenance of soil pH and the availability of organic carbon, nitrogen, phosphorus, potassium, calcium and magnesium in the soil may be influenced by shade tree interactions. For example, leguminous shade trees may play a positive role in fixing nitrogen (N), and hence the availability of nitrogen for the cocoa plants. Shade trees also contribute to nutrient recycling through litter decomposition (Asigbaase et al., 2021) and play a role in preventing soil erosion and regulating atmospheric temperatures that are directly linked to soil temperatures, moisture and cocoa root-associated microbiome and activities (Schmidt et al., 2022). Furthermore, shade trees provide habitats for fauna (birds, insects, etc.) that are essential in cocoa pollination and in the provision of other essential ecosystem services that benefit cocoa plants. However, the role of shade trees in cocoa systems depends on their general structural architecture above ground and whether the root systems overlap with the desired crops (cocoa) (Asante et al., 2021; Rigal et al., 2022).

Higher cocoa yield was reported under no-shade conditions compared to shaded conditions in a pioneering long-term study of the relation between shade and cocoa nutrition (Ahenkorah et al., 1987). This study was conducted with high inputs of fertilizer and other agrochemicals. These findings contrasted with our findings from the Western region of Ghana, where we observed more than 40% higher yield in shaded plots compared to unshaded plots (Asitoakor et al., 2022a). A main difference between the two studies, which may have led to the contrasting results, was that in our study agricultural inputs were relatively low, reflecting the input use of the majority of Ghanaian farmers. Our study also suggested modest impacts of shade trees on nutrient availability, as we observed no significant differences between shaded and unshaded plots in terms of soil concentrations of total nitrogen, exchangeable potassium, calcium and magnesium. Nonetheless, we found differences in the potentials of eight common forest shade tree species with regard to the concentration of available phosphorus (P) in comparison with the unshaded control plots (Fig. 3.1(a)) (Asitoakor et al., 2022a). The unshaded control plots in Fig. 3.1(a) showed the highest concentration of available P compared with plots with shade trees. The possibility that the shade trees may have competed with the cocoa plants and absorbed some of the soil P has been raised as a concern by some researchers (Gateau, 2018). Although there is less available soil P below shade tree species than in the control plots, this was not the limiting factor, as cocoa yield under these tree species were higher than in the unshaded control plots (Fig. 3.1(b)). This was expected, as Isaac et al. (2007) and Asare et al. (2017) have documented the possibility of improving yield from enhanced nutrient uptake by cocoa trees under shade trees when water is not a limiting factor.

Traditionally, cocoa farmers have sustained cocoa production through expansion into forest areas and/or by intensification through the addition of fertilizers (organic and/or inorganic) and through the chemical control of pests and diseases in varying quantities based on their affordability

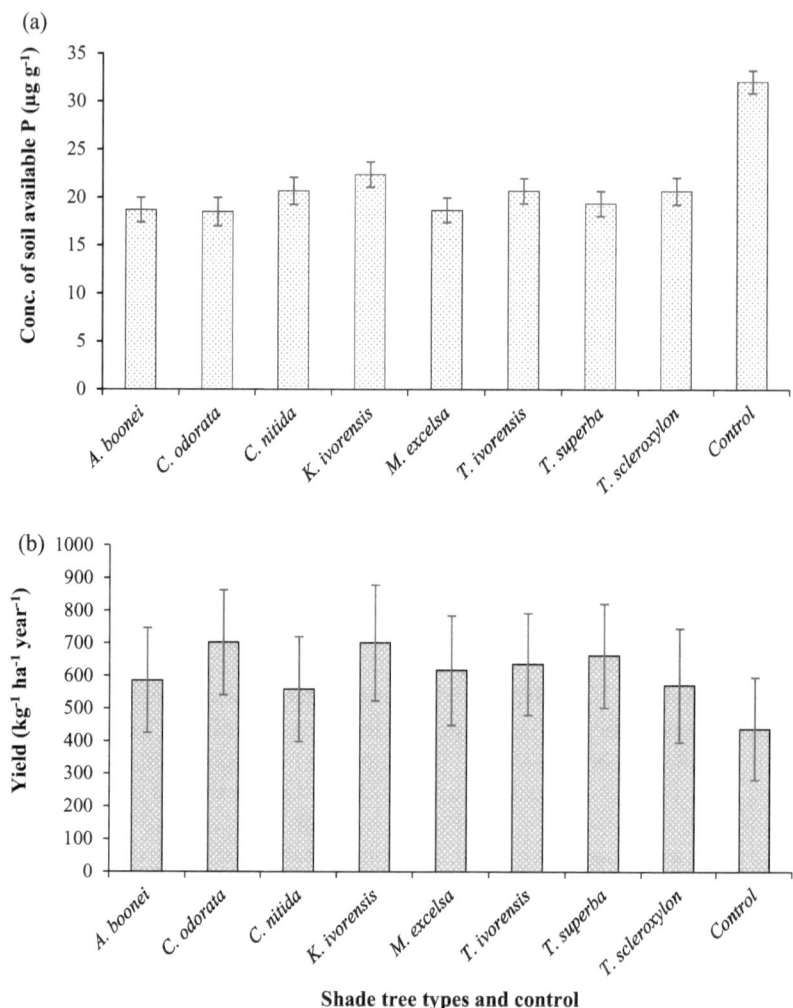

Fig. 3.1 The relationship between the selected shade tree species and unshaded control and **(a)** soil available phosphorus, and **(b)** cocoa yield in the three-year (2018–2021) field study (Values represent mean ± s.e. Creative Commons Attribution BY 4.0)

and availability. While expansion into forests results in forest degradation and loss of biodiversity, the irrational application of agro-chemical inputs could lead to soil degradation and other environmental problems such as water pollution, habitat destruction and biodiversity loss including beneficial pollinators (Adu-Acheampong et al., 2015; Bhandari, 2014). Although application of the right fertilizers increases yield (Asare et al., 2019), it could have negative influences on the composition of the root-associated microbiome of cocoa if the right amounts are not applied at the recommended rates, thereby reducing the decomposition of organic matter and the natural nutrient-recycling potential of cocoa soils (Niether et al., 2019; Schmidt et al., 2022). Likewise, the application of ammonium-based fertilizers can increase the acidity of cocoa soils and may reduce the potential of the soil to support yield after prolonged usage. Although organic fertilizers are considered better than inorganic fertilizers from an environmental safety perspective, few farmers use them. Since both organic and inorganic fertilizers are costly, planting and managing shade tree species that are known to improve soil fertility may be an economic alternative. There is a need for better management practices and policy incentives that reduce production costs, protect the environment and sustain yield.

Many studies have evaluated cocoa yield under both shaded and unshaded (full-sun) systems (Abdulai et al., 2018; Ahenkorah et al., 1987; Asare et al., 2017), but species-specific studies are rare. In our field study involving eight forest shade tree species, species such as *C. odorata*, *T. superba* and *K. ivorensis* resulted in significantly higher cocoa yield than full-sun control plots (Fig. 3.1(b)) (Asitoakor et al., 2022a). As mentioned above, the recorded yield did not correlate with nutrient availability (such as P) expressed by the control plots in Fig. 3.1(a). This showed that the productivity of cocoa is influenced by other factors than just soil fertility. This may include the architecture of the shade trees above the cocoa trees. The tree species with the highest yield, *C. odorata*, *T. superba* and *K. ivorensis*, all have tall stems and less dense canopies compared to species, such as *C. nitida*, which have relatively short stems and dense canopies. However, since there were no significant differences between species, further studies are needed before a definite conclusion can be made on this aspect. Asante et al. (2021) suggested that the architecture of shade trees is critical for levels of aeration, light penetration and the nutrient-recycling potential in cocoa-agroforestry systems. Interestingly, the average yield recorded in Fig. 3.1(b) in the plots under shade trees was higher than the unshaded control plots, as well as the national average cocoa yield between 400 and 550 kg ha^{-1} across Ghana, Côte d'Ivoire, Nigeria, Cameroon and Togo (Bymolt et al., 2018; Oomes et al., 2016).

3.3   SHADE TREE INFLUENCES ON COCOA ON-FARM PESTS AND DISEASES

Pests and diseases in cocoa are managed mainly by pesticide and fungicide applications. Mirids (*S. singularis* Hagl. and *D. theobroma* Dist.) and black pod diseases (caused by *P. palmivora* and *P. megakarya*) are the major cocoa pests and diseases in West Africa (Adu-Acheampong et al., 2015; Akrofi et al., 2015). Due to the environmental and health risks associated with the application of pesticides, coupled with the high costs involved, integrated pest management (IPM) approaches have been recommended to control pests and diseases in cocoa (Adu-Acheampong et al., 2015; Dormon et al., 2007). Integrated pest management relies on close monitoring and knowledge of the pests and pathogens and involves combining natural or biological pest control mechanisms (Bajwa & Kogan, 2002; Kabir & Rainis, 2015). As in other agricultural systems, the micro-climatic conditions (temperature, rainfall and relative humidity) strongly influence pest and disease occurrence and impact (De Almeida & Valle, 2007). For example, low temperatures and high relative humidity under shade trees favour black pod disease (Fig. 3.2(a)), while high temperatures under low rainfall conditions tend to favour some insects (e.g. mirid in Fig. 3.2(b)) (Abdulai et al., 2020; Dormon et al., 2007). In Africa, mirids are widespread in the major cocoa areas and cause up to 75% yield losses when uncontrolled (Anikwe et al., 2009; Padi, 1997). Black pod disease predominates on West African cocoa fields, resulting in up to 80% losses in cocoa yield (Akrofi et al., 2015). High relative humidity from high rainfall and poor drainage conditions in cocoa systems promote fungal black pod disease, which peaks in May–June on most cocoa farms in West Africa (Akrofi et al., 2015; Opoku et al., 2000). These occurrences may be minimized or regulated through the adoption and good management of agroforestry practices, including regular pruning and the removal of mistletoe and diseased pods to sustain cocoa yield.

In Ghana, governmental and non-governmental agencies organize farmer training, support extension services and provide support with pesticides, spraying and pruning programmes, all to reduce mirid infections and black pod disease (Baah & Anchirinah, 2011; Cocoa Health and Extension Division [CHED] & World Cocoa Foundation [WCF], 2016). Such efforts have been ongoing for decades (Adu-Acheampong et al.,

Fig. 3.2 Important biotic stressors in cocoa. (**a**) Pods infected by black pod disease (caused by *Phytophthora* sp.), (**b**) mirid insect (*Sahlbergella singularis*) and (**c**) damage symptoms of mirid infection on cocoa pods (*Source* (**a**) and (**c**): Asitoakor (2021); (**b**): Photo by Bawa Abuu)

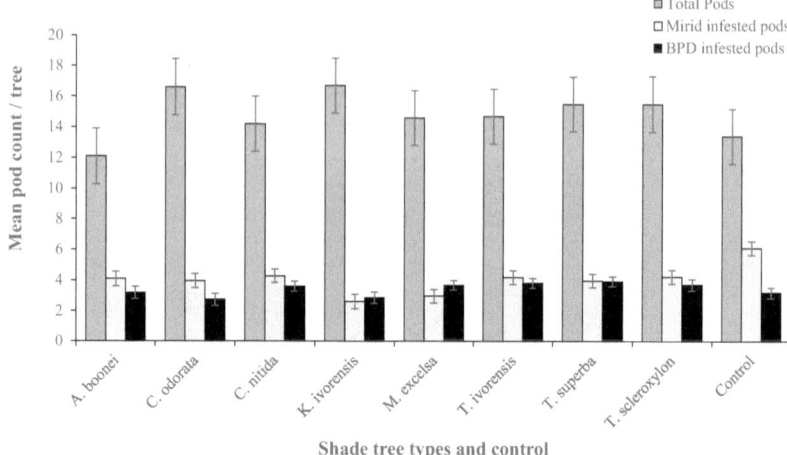

Fig. 3.3 Mean distribution of mirid insects and black pod disease infestations in cocoa pods under eight selected shade trees and in unshaded control areas (Values indicate mean ± s.e. Creative Commons Attribution BY 4.0)

2015; Akrofi et al., 2015). Since the 1950s, the development of resistant cocoa varieties and improved pesticide applications have been the major approaches to resolving the challenges of mirids and black pod diseases (Adu-Acheampong et al., 2015). However, the high costs of approved chemicals make farmers resort to using cheaper and unapproved pesticides that may reduce yield and risk contaminating the cocoa beans (Adu-Acheampong et al., 2015). With recent increases in global demand for cocoa beans free from agro-chemical residues (Cocoa Barometer, 2022), more biological ways of controlling pests and diseases in cocoa production have become desirable. Adu-Acheampong et al. (2015), however, question the existence of viable natural approaches to the management of cocoa pests and diseases. In our recent study of mirid insects and black pod disease infestation and damages in cocoa-agroforestry systems in Ghana, we observed variations in the magnitude of infection among eight selected shade trees species, as shown in Fig. 3.3 (Asitoakor et al., 2022a). The study confirmed that the level of mirid damage on cocoa pods may be effectively managed by the right combination of shade tree species and cocoa. Likewise, although black pod disease infections seem to be higher under shade trees than in unshaded plots, there could also be species-specific responses in that respect. The authors recommended further research to unravel species-specific responses to black pod infection in cocoa.

3.4 CONCLUSION AND POLICY IMPLICATIONS

In Ghana, shade trees are important components of cocoa-agroforestry systems, because they influence the occurrence and management of pests (mirids) and diseases (black pod disease) and improve yield in comparison to unshaded conditions. Shade trees have varying architecture, leaf sizes and crown densities that influence microclimate differently, thus impacting the occurrences of mirids and black pod diseases. Species such as Spanish cedar (*C. odorata*), limba (*T. superba*) and mahogany (*K. ivorensis*), which have less dense canopies, increase yield when used as shade trees in cocoa-agroforestry systems. Other species may have similar functions, but more research is required to understand how different shade tree species affect cocoa trees. Cocoa farmers are knowledgeable and have their preferred shade trees, and there is a need to combine local knowledge with scientific knowledge to guide the selection of shade tree species in cocoa-agroforestry systems to increase yield and mitigate climate change.

Current cocoa-related policies in Ghana promote the adoption of shade trees on cocoa farms with limited and unclear directions for selecting the specific types of shade trees. Unsuitable combinations of shade trees and cocoa may lead to increases in pest and disease incidence and severity, and negatively affect cocoa yield and quality. To ensure that the integration of shade trees does not harm cocoa production, and to increase farmers' interest in and satisfaction with keeping shade trees on their cocoa farms, policy development and dissemination by all relevant stakeholders in the cocoa sector is necessary. This includes the agricultural and cocoa-governing bodies such as the Ministry of Food and Agriculture of Ghana (MoFA) and the Ghana Cocoa Board (COCOBOD). Also private actors such as cocoa-based non-governmental organizations, farmer associations and cocoa-buying companies should critically consider the type of shade trees they recommend to cocoa farmers. Our study provides some insights on shade trees and the management of pests and diseases, but more knowledge is needed regarding the services and disservices of shade trees in cocoa cultivation.

Appendix: List of Common Shade Tree Species Adopted in Cocoa-Agroforestry Systems and Their Additional Uses

No	Scientific name	Common uses	Reference
1	*Albizia adianthifolia*	Treating diabetes, headache, eye problems, wounds, pain, skin diseases, gastrointestinal problems, haemorrhoids, infertility in women, respiratory problems and sexually transmitted infections	Lemmens (2007a)
2	*Albizia ferruginea*	Construction, flooring, staircases, furniture, cabinetry, joinery, turnery, carvings and veneer	Twum-Ampofo (2007)
3	*Albizia glaberrima*	Construction and furniture, stools, beehives, tool handles and grain mortars	Lemmens (2007b)
4	*Albizia zygia*	Carving, flooring and furniture. Bark decoction: treating bronchial diseases, fever, malaria, female sterility and as a purgative, stomachic, antidote, vermifuge and aphrodisiac	Apetorgbor (2007)
5	*Alstonia boonei*[a]	Boats, furniture, sculptures, musical instruments and firewood. Bark decoction: treating fractures and dislocations, jaundice and inducing breast milk	Adotey et al. (2012)
6	*Amphimas pterocarpoides*	Wood: interior construction, flooring, interior trim, joinery, furniture, canoes, huts. Bark decoction: treating dysentery, anaemia, haematuria, dysmenorrhoea, blennorrhoea, schistosomiasis, mumps and as a poison antidote	Tchinda and Tané (2008)

(continued)

(continued)

No	Scientific name	Common uses	Reference
7	*Anthocleista sp.*	Treating diabetes, hypertension, malaria, typhoid fever, obesity, diarrhoea, dysentery, abdominal and chest pain, ulcers, jaundice, asthma, haemorrhoids, hernia, cancer, rheumatism, STDs, infertility and skin diseases	Anyanwu et al. (2015)
8	*Antiaris toxicaria*	Sap: as an agent for immobilizing animals during hunting	Bosu and Krampah (2005b)
9	*Antrocaryon micraster*	Bark: preparing soup, treatment of malaria and as an enema to treat impotence and threatened abortion	Ayarkwa (2011)
10	*Blighia sapida*	Food and cosmetics production. Also for treating backache, constipation, cancer, fever, gonorrhoea, dysentery, psychosis, hernia, stomach-ache, malaria, rheumatism, typhoid, etc.	Sinmisola et al. (2019) and Asamoah et al. (2010)
11	*Bombax buonopozense*	Bark, flowers and leaves: treating ringworm, swellings, fever, convulsions and insanity and to clean hairy leather	Danso et al. (2019)
12	*Cedrela odorata*[a]	Cigar boxes, construction, joinery, mouldings, panelling, louvred doors, boats, furniture, cabinetry, household implements, musical instruments, carvings, veneer, plywood and turnery. Root and trunk bark: treating fever and pain	Lemmens (2008)
13	*Ceiba pentandra*	Construction. Its fluffy cotton-like seed pods are used as stuffing materials for cushions, pillows, mattresses, insulation and absorbent	Duvall (2011)
14	*Celtis mildbraedii*	Construction, furniture and ladders. Also for poles, pestles, tool handles and spoons	Oyen (2012)

(continued)

(continued)

No	Scientific name	Common uses	Reference
15	*Celtis zenkeri*	Construction, flooring and fuelwood	Essien and Oteng-Amoako (2012)
16	*Citrus sinensis*	Food and for producing beverages. The peel: increase appetite, reduce phlegm and treat coughs, colds, intestinal gas (flatulence) and acid indigestion	Yerou et al. (2017)
17	*Cola gigantea*	Treating sores, skin infections and pains	Atolani et al. (2019)
18	*Cola nitida*[a]	Producing beverages e.g. Coca Cola and Pepsi Cola. Nuts: the short-term relief of fatigue, depression, chronic fatigue syndrome (CFS) and melancholy	Atolani et al. (2019)
19	*Daniellia ogea*	Construction, flooring, joinery, furniture, novelties, boxes, crates, agricultural implements	Schmelzer (2012)
20	*Dialium dinklagei*	Food, medicine and as a source of wood	Lemmens (2012)
21	*Discoglypremna caloneura*	Bark decoction: for cough relief and intestinal pain from food poisoning. Bark powder: treating sores	Schmelzer (2008)
22	*Distemonanthus benthamianus*	Treating diarrheal infections and as wood for construction	Owusu and Louppe (2012)
23	*Entandrophragma angolense*	Bark: treating fever, stomach pain, peptic ulcers, earache, arthritic or rheumatic pain, swellings and ophthalmia, etc.	Tchinda (2008)
24	*Entandrophragma cylindricum*	Bark: treating bronchitis, lung complaints, colds, oedema, wounds and as an anodyne	Kémeuzé (2008)
25	*Erythrina vogelli*	Wood: floats for fishing nets and brake blocks and shingles. Branches: fence posts and for the relief of pain	Lemmens (2008)

(continued)

(continued)

No	Scientific name	Common uses	Reference
26	*Ficus capensis*	Leaf decoction: as fertility agent in men and for treating dysentery, oedema, leprosy, epilepsy, rickets, gonorrhoea, anaemia, tuberculosis and pains	Nworu et al. (2013)
27	*Ficus exasperata*	Root decoctions: treating urinary tract ailments, gonorrhoea, asthma and tuberculosis. Leaves: treating swellings, wounds and arthritic joints	Nworu et al. (2013)
28	*Ficus sur*	Ornamental use and hedges	Lumbile and Mogotsi (2008)
29	*Funtumia elastica*	Treating whooping cough, asthma, blennorrhoea, painful menstruation, fungal infections and wounds	Agyare et al. (2013)
30	*Glyphea brevis*	Treating fever, gonorrhoea, dysentery, stomach and lung troubles, parasitic infections, convulsions and constipation	Dickson et al. (2011)
31	*Gmelina arboria*	Construction, carving, musical instruments, pulp, particle board, plywood, matches and packing. Leaves: fodder and for rearing silkworms. Treating common cold, sore throat, cough and flu	Adam and Krampah (2005)
32	*Hannoa klaineana*	Treating fevers, malaria and gastrointestinal disorders	Abubakar et al. (2020)
33	*Holarrhena floribunda*	Wood: carvings, combs, spoons and handles for axes and small implements. Leaves: treating diabetes, malaria, cancer and oxidant damage dysentery, diarrhoea, fever, snakebite, infertility venereal disease	Schmelzer (2006)
34	*Iryingia gabonensis*	Wood: making utensils. Fruit: food and for weight loss, high cholesterol and diabetes	Mateus-Reguengo et al. (2019)

(continued)

(continued)

No	Scientific name	Common uses	Reference
35	*Khaya ivorensis*[a]	Wood: dugout canoes. Bark decoctions: treating coughs, fever, malaria, anaemia, wounds, sores, ulcers, tumours, rheumatic pains and lumbago. Root pulp is applied as an enema to treat dysentery	Lemmens (2008)
36	*Klainedoxa gabonensis*	Bark: treating rheumatism, lumbago, smallpox, chickenpox, fractures, dental caries, sterility and impotence	Oteng-Amoako and Obeng (2012)
37	*Lannea welwitschii*	Wood: furniture and utensils. Fruits: food. Bark; produce dye, make rope and treat diarrhoea, haemorrhoids, sterility of women, menstrual troubles, pain after childbirth, gonorrhoea, epilepsy, oedema, palpitation, skin infections and ulcers	Ebanyenle (2009)
38	*Lonchocarpus sericeus*	Remedy for pain and inflammation and as fuelwood	Amegnona and Messanvi (2009)
39	*Mangifera indica*	Food and as a beverage. Used as a dentifrice, antiseptic, astringent, diaphoretic, stomachic, vermifuge, tonic, laxative and diuretic and to treat diarrhoea, dysentery, anaemia, asthma, bronchitis, coughs, hypertension, insomnia, rheumatism, toothache, leucorrhoea, haemorrhage and piles. It is used as animal feed, fodder and forage	Lauricella et al. (2017)
40	*Margaritaria discoidea*	Bark: a purgative and for treating stomach-ache, toothache, post-partum pains, stomach and kidney complaints and to facilitate parturition. Wood: for poles, planks and shingles in housebuilding, flooring and interior trim	Addo-Danso (2012)

(continued)

(continued)

No	Scientific name	Common uses	Reference
41	*Milicia excelsa*[a]	Wood: construction, furniture, joinery, panelling, floors and boats/shipbuilding and marine carpentry, sleepers, sluice gates, framework, trucks, draining boards, outdoor and indoor joinery. Bark: treating cough, asthma, heart trouble, lumbago, spleen pain, stomach pain, abdominal pain, oedema, ascites, dysmenorrhoea, gonorrhoea, general fatigue, rheumatism, sprains and as a galactagogue, aphrodisiac, tonic and purgative, treatment of snakebites and fever	Ofori (2007)
42	*Morinda lucida*	Bark, leaves and roots: treating malaria, diabetes, hypertension, inflammation, typhoid fever, cancer, cognitive disorders, sickle cell disease, trypanosomiasis, onchocerciasis and irregular menstruation, insomnia, wounds infections and jaundice	Abbiw (1990) and Zimudzi and Cardon (2005)
43	*Morus mesozygia*	All plant parts: in decoctions, baths, massages and enemas as treatments for rheumatism, lumbago, intercostal pain, neuralgia, colic, stiffness, debility, diarrhoea and dysentery. The root: as an aphrodisiac	Toirambe Bamoninga and Ouattara (2008)

(continued)

(continued)

No	Scientific name	Common uses	Reference
44	*Musanga cecropioides*	Stem sap: treating dysmenorrhoea and galactagogue. Root sap: treating stomach spasms, diarrhoea, gonorrhoea, pulmonary complaints, trypanosomiasis, skin diseases, otitis, rheumatism, oedema, epilepsy and to ease childbirth. Wood: interior construction. Bark: treating chest pains	Todou and Meikeu Kamdem (2011)
45	*Nesogordonia papaverifera*	Wood: exterior and interior joinery, parquetry, turnery, staircase boards, window frames, furniture, cabinets, tool handles, mallets, lorry bodies, coach/wagon work and small boats, carving, sliced veneer, plywood and firewood. Leaf decoction: dental caries relief	Oyen (2005)
46	*Newbouldia laevis*	Treating coughs, malaria, diarrhoea, elephantiasis, epilepsy and dysentery, epilepsy and convulsions in children. Bark: as enema for treating constipation and piles, septic wounds and as firewood	Dermane et al. (2020)
47	*Pentaclethra macrophylla*	Leaf, bark, seed extracts and fruit pulp: treating gonorrhoea and convulsions. Also as an analgesic, laxative, enema against dysentery and liniment against itch. As firewood and charcoal	Oboh (2007)
48	*Persea americana*	Leaves: treating dysentery, coughs, high blood pressure, liver problems and gout. Bark: treating diarrhoea, fruits for lowering blood cholesterol level, promote hair growth and to treat skin conditions. It is also used to boost sexual longing	Tcheghebe et al. (2016)

(continued)

(continued)

No	Scientific name	Common uses	Reference
49	*Petersianthus macrocarpus*	Wood: construction, furniture, canoes, mortars, tool handles, sliced veneer and plywood, flooring, mine props, vehicle bodies, railway sleepers, sporting goods, toys, novelties, agricultural implements and draining boards. Treating pains, headaches and fever	Owusu (2012)
50	*Psidium guajava*	Treating inflammation, diabetes, hypertension, dysentery, caries, wounds, pain relief, fever, diarrhoea, rheumatism, lung diseases and ulcers	Daswani et al. (2017)
51	*Pterygota macrocarpa*	Treating sores, skin infections, stomach-ache, digestive disorders and pains. Wood: veneer, plywood, interior panelling, interior joinery, moulding, furniture and block board	Oyen (2008)
52	*Pycnanthus angolensis*	Bark: poison antidote and for treating leprosy, anaemia, infertility, gonorrhoea and malaria. Leaf extracts: for enema to treat oedema. Root extracts: treating schistosomiasis. As purgative and for cleansing milk of lactating mothers and for the treating coughs and chest pains	Mapongmetsem (2007)
53	*Ricinodendron heudelotti*	Bark: treating gonorrhoea, cough, leprosy, hernia, dysentery, elephantiasis, syphilis, yellow fever, anaemia, toothache and malaria. Wood: plywood for building and construction	Tchoundjeu and Atangana (2007)

(continued)

(continued)

No	Scientific name	Common uses	Reference
54	*Spathodea campanulata*	Food and for treating epilepsy and convulsion, kidney disease, urethritis, also as antidote for animal poisons, inflamed skin and rashes	Bosch (2002)
55	*Spondias mombin*	Treating diarrhoea, fracture, convulsion, wounds, eye and ringworm. The fruit is used for a juice drink. Leaf decoction: treating laryngitis, tooth decay, cough, sore throat and malaria	Nworu et al. (2011)
56	*Sterculia tragacantha*	Treating boils, diarrhoea, dyspepsia, fever, gonorrhoea, snake bite, syphilis and tapeworm and managing diabetes mellitus	Owusu and Derkyi (2011)
57	*Tectona grandis*	Oil extract: treating scabies and as hair tonic. Bark: treating bronchitis. Wood: construction and poles	Louppe (2005)
58	*Terminalia ivorensis*[a]	Treating dermal diseases, for firewood and charcoal. Wood: joinery, cabinetry and furniture	Foli (2009)
59	*Terminalia superba*[a]	Bark decoctions: treating wounds, sores, haemorrhoids, diarrhoea, dysentery, malaria, vomiting, gingivitis, bronchitis, aphthae, swellings, ovarian troubles, diabetes mellitus, gastroenteritis and jaundice. Wood: furniture, table tennis boards	Kimpouni (2009)
60	*Trema orientalis*	Leaves and bark: gargling, inhalation, drink, lotion, bath or vapour baths for coughs, sore throat, asthma, bronchitis, gonorrhoea coughs, yellow fever, toothache and as an antidote to general poisoning. Wood: construction, firewood and charcoal	Orwa et al. (2009)

(continued)

(continued)

No	Scientific name	Common uses	Reference
61	*Trichilia manodelpha*	Treating epilepsy, depression, pain and psychosis and inflammatory conditions rheumatism, oedema, gout. Also used as firewood and charcoal	Lemmens (2008)
62	*Trilepisium madagascariense*	Leaves are used as vegetables and other parts for treating pain and venereal diseases	Ango et al. (2012)
63	*Triplochiton scleroxylon*[a]	Its sawdust is used in raising edible fungi (Pleurotus spp). Bark: to cover the roof and walls of huts. Wood: fibreboard, fuelwood and carving	Bosu and Krampah (2005a)
64	*Zanthoxylum gilletii*	Bark of stem and roots: treating burns, rheumatism, headache, stomach-ache, toothache and pain after childbirth. Bark: against colic, fever and in managing malaria, tumours and sickle cell anaemia	Okagu et al. (2021)

Note Shade tree species were selected based on findings from Asare (2016) and Graefe et al. (2017)
[a]Species used or assessed in this study

References

Abbiw, D. K. (1990). *Useful plants of Ghana: West African uses of wild and cultivated plants* (337 pp.). Intermediate Technology Publications.

Abdulai, I., Hoffmann, M. P., Jassogne, L., Asare, R., Graefe, S., Tao, H. H., Muilerman, S., Vaast, P., Van Asten, P., Läderach, P., & Rötter, R. P. (2020, March). Variations in yield gaps of smallholder cocoa systems and the main determining factors along a climate gradient in Ghana. *Agricultural Systems, 181*, 1–8. https://doi.org/10.1016/j.agsy.2020.102812

Abdulai, I., Jassogne, L., Graefe, S., Asare, R., Van Asten, P., Läderach, P., & Vaast, P. (2018). Characterization of cocoa production, income diversification and shade tree management along a climate gradient in Ghana. *PLoS ONE, 13*(4), 1–17. https://doi.org/10.1371/journal.pone.0195777

Abubakar, I., Yankuzo, H., Shuaibu, M. Y. B., & Abubakar, M. G. (2020). Anti-ulcer activity of methanol extract of the leaves of Hannoa klaineana in rats. *The Journal of Phytopharmacology, 9*(4), 258–264.

Adam, K. A., & Krampah, E. (2005). Gmelina arborea Roxb. ex Sm. In D. Louppe, A. A. Oteng-Amoako, & M. Brink (Eds.), *PROTA (Plant Resources of Tropical Africa/Ressources végétales de l'Afrique tropicale).* Wageningen University. Accessed 20 January 2023.

Addo-Danso, A. (2012). Margaritaria discoidea (Baill.) G. L. Webster. In R. H. M. J. Lemmens, D. Louppe, & A. A. Oteng-Amoako (Eds.), *PROTA (Plant Resources of Tropical Africa/Ressources végétales de l'Afrique tropicale).* Wageningen University. Retrieved on 20 January 2023, from http://www.prota4u.org/search.asp

Adjaloo, M. K., Oduro, W., & Banful, B. K. (2012). Floral phenology of upper Amazon cocoa trees: Implications for reproduction and productivity of cocoa. *ISRN Agronomy, 2012,* 1–8. https://doi.org/10.5402/2012/461674

Adotey J. P., Adukpo G. E., Opoku Boahen Y., & Armah, F. A. (2012). A review of the ethnobotany and pharmacological importance of Alstonia boonei De Wild (Apocynaceae). *ISRN Pharmacology,* 587160. https://doi.org/10.5402/2012/587160

Adu-Acheampong, R., Sarfo, J., Appiah, E., Nkansah, A., Awudzi, G., Obeng, E., Tagbor, P., & Sem, R. (2015). Strategy for insect pest control in cocoa. *American Journal of Experimental Agriculture, 6*(6), 416–423. https://doi.org/10.9734/ajea/2015/12956

Agyare, C., Koffuor, G. A., Boakye, Y. D., & Mensah, K. B. (2013). Antimicrobial and anti-inflammatory properties of Funtumia elastica. *Pharmaceutical Biology, 51*(4), 418–425. https://doi.org/10.3109/13880209.2012.738330

Ahenkorah, Y., Halm, B. J., Appiah, M. R., Akrofi, G. S., & Yirenkyi, J. E. K. (1987). Twenty years' results from a shade and fertilizer trial on Amazon cocoa (Theobroma cacao) in Ghana. *Experimental Agriculture, 23,* 31–39.

Akrofi, A. Y., Amoako-Atta, I., Assuah, M., & Asare, E. K. (2015). Black pod disease on cacao (Theobroma cacao, L) in Ghana: Spread of Phytophthora megakarya and role of economic plants in the disease epidemiology. *Crop Protection, 72,* 66–75. https://doi.org/10.1016/j.cropro.2015.01.015

Amegnona, A., & Mess Anvi, G. (2009). Hepatoprotective effect of Lonchocarpus sericeus leaves in CCl4-induced liver damage. *Journal of Herbs, Spices & Medicinal Plants, 15*(2), 216–226. https://doi.org/10.1080/10496470903139512

Ango, P. Y., Kapche, D. W. F. G., Kuete, V., Ngadjui, B. T., Bezabih, M., & Abegaz, B. M. (2012). Chemical constituents of *Trilepisium madagascariense* (*Moraceae*) and their antimicrobial activity. *Phytochemistry Letters, 5*(3), 524–528.

Anikwe, J. C., Omoloye, A. A., Aikpokpodion, P. O., Okelana, F. A., & Eskes, A. B. (2009). Evaluation of resistance in selected cocoa genotypes to the brown cocoa mirid, Sahlbergella singularis Haglund in Nigeria. *Crop Protection, 28*(4), 350–355. https://doi.org/10.1016/j.cropro.2008.11.014

Anyanwu, G. O., Rehman, N., Onyeneke, C. E., & Rauf, K. (2015). Medicinal plants of the genus Anthocleista. A review of their ethnobotany, phytochemistry and pharmacology. *Journal of Ethnopharmacology, 175*, 648–667.

Apetorgbor, M. M. (2007). Albizia zygia (DC.) J. F. Macbr. In D. Louppe, A. A. Oteng-Amoako, & M. Brink (Eds.), *PROTA (Plant Resources of Tropical Africa/Ressources végétales de l'Afrique tropicale)*. Wageningen University. Accessed 19 January 2023.

Asamoah, A., Antwi-Bosiako, C., Frimpong-Mensah, K., Atta-Boateng, A., Montes, C. S., & Louppe, D. (2010). Blighia sapida K. D. Koenig. [Internet] Record from PROTA4U. In R. H. M. J. Lemmens, D. Louppe, & A. A. Oteng-Amoako (Eds.), *PROTA (Plant Resources of Tropical Africa/Ressources végétales de l'Afrique tropicale)*. Wageningen University. Retrieved on 20 January 2023, from http://www.prota4u.org/search.asp

Asante, P. A., Rozendaal, M. A., Rahn, E., Zuidema, P. A., Quaye, A. K., Asare, R., Peter, L., & Anten, N. P. R. (2021). Unravelling drivers of high variability of on-farm cocoa yields across environmental gradients in Ghana. *Agricultural Systems, 193*, 1–10.

Asante-Poku, A., & Angelucci, F. (2013). *Analysis of incentives and disincentives for cocoa in Ghana* (Technical notes series; Issue June). MAFAP, FAO. http://www.fao.org/3/a-at593e.pdf

Asare, R. (2016). *The relationships between on-farm shade trees and cocoa yields in Ghana* (IGN PhD thesis November 2015). Department of Geosciences and Natural Resource Management, University of Copenhagen, Frederiksberg, pp. 1–46.

Asare, R., Asare, R. A., Asante, W. A., Markussen, B., & Raebild, A. (2017). Influences of shading and fertilization on on-farm yields of cocoa in Ghana. *Experimental Agriculture*, 1–16. https://doi.org/10.1017/S0014479716000466

Asare, R., Markussen, B., Asare, R. A., Anim-Kwapong, G., & Ræbild, A. (2019). On-farm cocoa yields increase with canopy cover of shade trees in two agro-ecological zones in Ghana. *Climate and Development, 11*(5), 1–12. https://doi.org/10.1080/17565529.2018.1442805

Asigbaase, M., Dawoe, E., Lomax, B. H., & Sjogersten, S. (2021, March). Biomass and carbon stocks of organic and conventional cocoa agroforests, Ghana. *Agriculture, Ecosystems and Environment, 306*, 1–11. https://doi.org/10.1016/j.agee.2020.107192

Asitoakor, B. K. (2021). *Effects of agroforestry and climate on cocoa yield*. University of Ghana.

Asitoakor, B. K., Asare, R., Ræbild, A., Ravn, H. P., Eziah, V. Y., Owusu, K., Mensah, E. O., & Vaast, P. (2022a). Influences of climate variability on cocoa health and productivity in agroforestry systems in Ghana. *Agricultural and Forest Meteorology, 327*(109199), 1–13. https://doi.org/10.1016/j.agrformet.2022.109199

Asitoakor, B. K., Vaast, P., Ræbild, A., Ravn, H. P., Eziah, V. Y., Owusu, K., Mensah, E. O., & Asare, R. (2022b). Selected shade tree species improved cocoa yields in low-input agroforestry systems in Ghana. *Agricultural Systems, 202*(103476), 1–9.

Atolani, O., Oguntoye, H., Areh, E. T., Adeyemi, O. S., & Kambizi, L. (2019). Chemical composition, anti-toxoplasma, cytotoxicity, antioxidant, and anti-inflammatory potentials of Cola gigantea seed oil. *Pharmaceutical Biology, 57*(1), 154–160. https://doi.org/10.1080/13880209.2019.1577468

Ayarkwa, J. (2011). Antrocaryon micraster A. Chev. & Guill. In R. H. M. J. Lemmens, D. Louppe, & A. A. Oteng-Amoako (Eds.), *PROTA (Plant Resources of Tropical Africa/Ressources végétales de l'Afrique tropicale).* Wageningen University. Accessed 20 January 2023.

Baah, F., & Anchirinah, V. (2011). A review of Cocoa Research Institute of Ghana extension activities and the management of cocoa pests and diseases in Ghana. *American Journal of Social and Management Sciences, 2*(1), 196–201. https://doi.org/10.5251/ajsms.2011.2.1.196.201

Babin, R., Gerben, M., Hoopen, T., Cilas, C., Enjalric, F., Yede, Gendre, P., & Lumaret, J. P. (2010). Impact of shade on the spatial distribution of Sahlbergella singularis in traditional cocoa agroforests. *Agricultural and Forest Entomology, 12*(1), 69–79. https://doi.org/10.1111/j.1461-9563.2009.00453.x

Bajwa, W. I., & Kogan, M. (2002). *Compendium of IPM Definitions (CID)* (Issue 998). IPPC Publication.

Bhandari, G. (2014). An overview of agrochemicals and their effects on environment in Nepal. *Applied Ecology and Environmental Sciences, 2*(2), 66–73. https://doi.org/10.12691/aees-2-2-5

Bosch, C. H. (2002). Spathodea campanulata P. Beauv. [Internet] Record from PROTA4U. In L. P. A. Oyen & R. H. M. J. Lemmens (Eds.), *PROTA (Plant Resources of Tropical Africa/Ressources végétales de l'Afrique tropicale).* Wageningen University. Accessed 20 January 2023.

Bosu, P. P., & Krampah, E. (2005a). Triplochiton scleroxylon K. Schum. In D. Louppe, A. A. Oteng-Amoako, & M. Brink (Eds.), *PROTA (Plant Resources of Tropical Africa/Ressources végétales de l'Afrique tropicale).* Wageningen University. Accessed 20 January 2023.

Bosu, P. P., & Krampah, E. (2005b). Antiaris toxicaria Lesch. In D. Louppe, A. A. Oteng-Amoako, & M. Brink (Eds.), *PROTA (Plant Resources of Tropical Africa/Ressources végétales de l'Afrique tropicale)*. Wageningen University. Accessed 20 January 2023.

Bymolt, R., Laven, A., & Tyszler, M. (2018). Production and yield. In *Demystifying the cocoa sector in Ghana and Côte d'Ivoire* (pp. 194–206). The Royal Tropical Institute (KIT). http://edepot.wur.nl/314177

Cocoa Barometer. (2022). *Cocoa Barometer, executive summary*. www.cocoabarometer.org

Cocoa Health and Extension Division [CHED], & World Cocoa Foundation [WCF]. (2016). *Manual for cocoa extension in Ghana*.

Danso, J., Alemawor, F., Boateng, R., Barimah, J., & Kumah, D. B. (2019). Effect of drying on the nutrient and anti-nutrient composition of Bombax buonopozense sepals. *African Journal of Food Science, 13*(1), 21–29.

Daswani, P. G., Gholkar, M. S., & Birdi, T. J. (2017). *Psidium guajava*: A single plant for multiple health problems of rural Indian population. *Pharmacognosy Reviews, 11*(22), 167–174. https://doi.org/10.4103/phrev.phrev_17_17

Daymond, A. J., & Hadley, P. (2008). Differential effects of temperature on fruit development and bean quality of contrasting genotypes of cacao (Theobroma cacao). *Annals of Applied Biology, 153*(2), 175–185. https://doi.org/10.1111/j.1744-7348.2008.00246.x

De Almeida, A. F., & Valle, R. R. (2007). Ecophysiology of the cacao tree. *Brazilian Journal of Plant Physiology, 19*(4), 425–448.

Dermane, A., Kpegba, K., Eloh, K., Osei-Safo, D., Amewu, R. K., & Caboni, P. (2020). Differential constituents in roots, stems and leaves of Newbouldia laevis Thunb. screened by LC/ESI-Q-TOF-MS. *Results in Chemistry, 2*, 100052.

Dickson, R. A., Ekuadzi, E., Annan, K., & Komlaga, G. (2011). Antibacterial, anti-inflammatory, and antioxidant effects of the leaves and stem bark of Glyphaea brevis (Spreng) Monachino (Tiliaceae): A comparative study. *Pharmacognosy Research, 3*(3), 166–172. https://doi.org/10.4103/0974-8490.85001

Dormon, E. N. A., Van Huis, A., & Leeuwis, C. (2007). Effectiveness and profitability of integrated pest management for improving yield on smallholder cocoa farms in Ghana. *International Journal of Tropical Insect Science, 27*(1), 27–39. https://doi.org/10.1017/S1742758407727418

Duvall, C. S. (2011). Ceiba pentandra (L.) Gaertn. In M. Brink & E. G. Achigan-Dako (Eds.), *PROTA (Plant Resources of Tropical Africa/Ressources végétales de l'Afrique tropicale)*. Wageningen University. Accessed 20 January 2023.

Ebanyenle, E. (2009). Lannea welwitschii (Hiern) Engl. In R. H. M. J. Lemmens, D. Louppe, & A. A. Oteng-Amoako (Eds.), *PROTA (Plant*

Resources of Tropical Africa/Ressources végétales de l'Afrique tropicale). Wageningen University. Accessed 20 January 2023.

Edwin, J., & Masters, W. A. (2005). Genetic improvement and cocoa yields in Ghana. *Experimental Agriculture, 41*(4), 491–503. https://doi.org/10. 1017/S0014479705002887

Essien, C., & Oteng-Amoako, A. A. (2012). Celtis zenkeri Engl. [Internet] Record from PROTA4U. In R. H. M. J. Lemmens, D. Louppe, & A. A. Oteng-Amoako (Eds.), *PROTA (Plant Resources of Tropical Africa/Ressources végétales de l'Afrique tropicale)*. Wageningen University. Accessed 20 January 2023.

Foli, E. G. (2009). Terminalia ivorensis A. Chev. In R. H. M. J. Lemmens, D. Louppe, & A. A. Oteng-Amoako (Eds.), *PROTA (Plant Resources of Tropical Africa/Ressources végétales de l'Afrique tropicale)*. Wageningen University. Accessed 20 January 2023.

Gateau, L. A. M. (2018, November). *Cocoa yield, nutrients and shade trees in traditional cocoa agroforests in a climate change context: A case study in Bahia, Brazil.*

Graefe, S., Meyer-Sand, L. F., Chauvette, K., Abdulai, I., Jassogne, L., Vaast, P., & Asare, R. (2017). Evaluating farmers' knowledge of shade trees in different cocoa agro-ecological zones in Ghana. *Human Ecology, 45*(3), 321–332. https://doi.org/10.1007/s10745-017-9899-0

Hoffmann, M. P., Cock, J., Samson, M., Janetski, N., Janetski, K., Rötter, R. P., Fisher, M., & Oberthür, T. (2020). Fertilizer management in smallholder cocoa farms of Indonesia under variable climate and market prices. *Agricultural Systems, 178*(February 2018), 1–13. https://doi.org/10.1016/j.agsy. 2019.102759

IITA. (2009). *"Climate change and cocoa": Annual report 2008/2009.*

Isaac, M., Timmer, V., & Quashie-Sam, S. (2007). Shade tree effects in an 8-year-old cocoa agroforestry system: Biomass and nutrient diagnosis of Theobroma cacao by vector analysis. *Nutrient Cycling in Agroecosystems, 78*(2), 155–165.

Kaba, J. S., Otu-nyanteh, A., & Abunyewa, A. A. (2020). The role of shade trees in influencing farmers' adoption of cocoa agroforestry systems: Insight from semi-deciduous rain forest agroecological zone of Ghana. *NJAS—Wageningen Journal of Life Sciences, 92*(100332), 1–7. https://doi.org/10.1016/j.njas. 2020.100332

Kabir, M. H., & Rainis, R. (2015). Do farmers not widely adopt environmentally friendly technologies? Lesson from Integrated Pest Management (IPM). *Modern Applied Science, 9*(3), 208–215. https://doi.org/10.5539/mas.v9n 3p208

Kémeuzé, V. A. (2008). Entandrophragma cylindricum (Sprague) Sprague. In D. Louppe, A. A. Oteng-Amoako, & M. Brink (Eds.), *PROTA (Plant Resources of Tropical Africa/Ressources végétales de l'Afrique tropicale)*. Wageningen University. Accessed 20 January 2023.

Kimpouni, V. (2009). Terminalia superba Engl. & Diels. In R. H. M. J. Lemmens, D. Louppe, & A. A. Oteng-Amoako (Eds.), *PROTA (Plant Resources of Tropical Africa/Ressources végétales de l'Afrique tropicale)*. Wageningen University. Accessed 20 January 2023.

Lauricella, M., Emanuele, S., Calvaruso, G., Giuliano, M., & D'Anneo, A. (2017) Multifaceted health benefits of Mangifera indica L. (Mango): The inestimable value of orchards recently planted in sicilian rural areas. *Nutrients.* 9(5), 525. https://doi.org/10.3390/nu9050525

Lemmens, R. H. M. J. (2007a). Albizia adianthifolia (Schumach.) W. Wight. In D. Louppe, A. A. Oteng-Amoako, & M. Brink (Eds.), *PROTA (Plant Resources of Tropical Africa/Ressources végétales de l'Afrique tropicale)*. Wageningen University. Accessed 19 January 2023.

Lemmens, R. H. M. J. (2007b). Albizia glaberrima (Schumach. & Thonn.) Benth. In D. Louppe, A. A. Oteng-Amoako, & M. Brink (Eds.), *PROTA (Plant Resources of Tropical Africa/Ressources végétales de l'Afrique tropicale)*. Wageningen University. Accessed 19 January 2023.

Lemmens, R. H. M. J. (2008). Cedrela odorata L. In D. Louppe, A. A. Oteng-Amoako, & M. Brink (Eds.), *PROTA (Plant Resources of Tropical Africa/ Ressources végétales de l'Afrique tropicale)*. Wageningen University. Accessed 20 January 2023.

Lemmens, R. H. M. J. (2012). Dialium aubrevillei Pellegr. [Internet] Record from PROTA4U. In R. H. M. J. Lemmens, D. Louppe, & A. A. Oteng-Amoako (Eds.), *PROTA (Plant Resources of Tropical Africa/Ressources végétales de l'Afrique tropicale)*. Wageningen University. Accessed 20 January 2023.

Louppe, D. (2005). Tectona grandis L.f. In D. Louppe, A. A. Oteng-Amoako, & M. Brink (Eds.), *PROTA (Plant Resources of Tropical Africa/Ressources végétales de l'Afrique tropicale)*. Wageningen University. Accessed 20 January 2023.

Lumbile, A. U., & Mogotsi, K. K. (2008). Ficus sur Forssk. In D. Louppe, A. A. Oteng-Amoako, & M. Brink (Eds.), *PROTA (Plant Resources of Tropical Africa/Ressources végétales de l'Afrique tropicale)*. Wageningen University. Accessed 20 January 2023.

Mahob, R. J., Baleba, L., Yede, Dibog, L., Cilas, C., Bilong Bilong, C. F., & Babin, R. (2015). Spatial distribution of Sahlbergella singularis hagl. (hemiptera:Miridae) populations and their damage in unshaded young cacao-based agroforestry systems. *International Journal of Plant, Animal and Environmental Sciences, 5*(2), 121–132.

Mapongmetsem, P. M. (2007). Pycnanthus angolensis (Welw.) Warb. In H. A. M. van der Vossen & G. S. Mkamilo (Eds.), *PROTA (Plant Resources of Tropical Africa/Ressources végétales de l'Afrique tropicale)*. Wageningen University. Accessed 20 January 2023.

Mateus-Reguengo, L., Barbosa-Pereira, L., Rembangouet, W., Bertolino, M., Giordano, M., Rojo-Poveda, O., & Zeppa, G. (2019). Food applications of *Irvingia gabonensis* (Aubry-Lecomte ex. O'Rorke) Baill., the 'bush mango': A review. *Critical Reviews in Food Science and Nutrition, 60*(14), 2446–2459. https://doi.org/10.1080/10408398.2019.1646704

Mcelroy, M. S., Navarro, A. J. R., Mustiga, G., Stack, C., Gezan, S., Sarabia, W., Saquicela, D., Sotomayor, I., Douglas, G. M., Amores, F., Tarqui, O., Myles, S., & Motamayor, J. C. (2018, March). Prediction of cacao (Theobroma cacao) resistance to Moniliophthora spp. diseases via genome-wide association analysis and genomic selection. *Frontiers in Plant Science, 9*, 1–12. https://doi.org/10.3389/fpls.2018.00343

Medina, V., & Laliberte, B. (2017). *A review of research on the effects of drought and temperature stress and increased CO_2 on Theobroma cacao L., and the role of genetic diversity to address climate change*. Bioversity International. https://www.bioversityinternational.org/fileadmin/user_upload/Review_laliberte_2017_new.pdf

Niether, W., Schneidewind, U., Fuchs, M., Schneider, M., & Armengot, L. (2019). Below- and aboveground production in cocoa monocultures and agroforestry systems. *Science of the Total Environment, 657*, 558–567. https://doi.org/10.1016/j.scitotenv.2018.12.050

Nworu, C. S., Akah, P. A., Okoye, F. B. C., Toukam, D. K., Udeh, J., & Esimone, C. O. (2011). The leaf extract of Spondias mombin L. displays an anti-inflammatory effect and suppresses inducible formation of tumor necrosis factor-α and nitric oxide (NO). *Journal of Immunotoxicology, 8*(1), 10–16. https://doi.org/10.3109/1547691X.2010.531406

Nworu, C. S., Nwuke, H. C., Akah, P. A., Okoye, F. B. C., & Esimone, C. O. (2013). Extracts of Ficus exasperata leaf inhibit topical and systemic inflammation in rodents and suppress LPS-induced expression of mediators of inflammation in macrophages. *Journal of Immunotoxicology, 10*(3), 302–310. https://doi.org/10.3109/1547691X.2012.732121

Oboh, G. (2007). Pentaclethra macrophylla Benth. [Internet] Record from PROTA4U. In H. A. M. van der Vossen & G. S. Mkamilo (Eds.), *PROTA (Plant Resources of Tropical Africa/Ressources végétales de l'Afrique tropicale)*. Wageningen University. Retrieved on 20 January 2023, from http://www.prota4u.org/search.asp

Ofori, D. A. (2007). Milicia excelsa (Welw.) C. C. Berg. In D. Louppe, A. A. Oteng-Amoako, & M. Brink (Eds.), *PROTA (Plant Resources of Tropical Africa/Ressources végétales de l'Afrique tropicale)*. Wageningen University. Accessed 20 January 2023.

Ofori-Frimpong, K., Asase, A., & Yelibora, M. (2007). *Cocoa farming and biodiversity in Ghana: Annual report 2007*.

Okagu, I. U., Ndefo, J. C., Aham, E. C., & Udenigwe, C. C. (2021). *Zanthoxylum* species: A review of traditional uses, phytochemistry and pharmacology in relation to cancer, infectious diseases and sickle cell anemia. *Frontiers in Pharmacology, 12*, 713090. https://doi.org/10.3389/fphar.2021.713090

Oomes, N., Tieben, B., Laven, A., Ammerlaan, T., Appleman, R., Biesenbeek, C., & Buunk, E. (2016). *Market concentration and price formation in the global cocoa value chain* (SEO-rapport; No. 2016-79). SEO Economisch Onderzoek. http://www.seo.nl/pagina/article/market-concentration-and-price-formation-in-the-global-cocoa-value-chain/

Opoku, I. Y., Appiah, A. A., Akrofi, A. Y., & Owusu, G. K. (2000). Phytophthora megakarya: A potential threat to the cocoa industry in Ghana. *Ghana Journal of Agricultural Science, 33*(2), 1–13. https://doi.org/10.4314/gjas.v33i2.1876

Orwa, C., Mutua, A., Kindt, R., Jamnadass, R, & Anthony, S. (2009). *Agroforestree database: A tree reference and selection guide version 4.0*. Retrieved on 20 January 2023 from, http://www.worldagroforestry.org/sites/treedbs/treedatabases.asp

Oteng-Amoako, A. A., & Obeng, E. A. (2012). Klainedoxa gabonensis Pierre. In R. H. M. J. Lemmens, D. Louppe, & A. A. Oteng-Amoako (Eds.), *PROTA (Plant Resources of Tropical Africa/Ressources végétales de l'Afrique tropicale)*. Wageningen University. Accessed 20 January 2023.

Owusu, F. W. (2012). Petersianthus macrocarpus (P. Beauv.) Liben. In R. H. M. J. Lemmens, D. Louppe, & A. A. Oteng-Amoako (Eds.), *PROTA (Plant Resources of Tropical Africa/Ressources végétales de l'Afrique tropicale)*. Wageningen University. Accessed 20 January 2023.

Owusu, F. W., & Derkyi, N. S. A. (2011). Sterculia africana (Lour.) Fiori. [Internet] Record from PROTA4U. In M. Brink & E. G. Achigan-Dako (Eds.), *PROTA (Plant Resources of Tropical Africa/Ressources végétales de l'Afrique tropicale)*. Wageningen University. Retrieved on 20 January 2023 from, http://www.prota4u.org/search.asp

Owusu, F. W., & Louppe, D. (2012). Distemonanthus benthamianus Baill. [Internet] Record from PROTA4U. In R. H. M. J. Lemmens, D. Louppe, & A. A. Oteng-Amoako (Eds.), *PROTA (Plant Resources of Tropical Africa/Ressources végétales de l'Afrique tropicale)*. Wageningen University.

Oyen, L. P. A. (2005). Nesogordonia kabingaensis (K. Schum.) Capuron ex R.Germ. In D. Louppe, A. A. Oteng-Amoako, & M. Brink (Eds.), *PROTA (Plant Resources of Tropical Africa/Ressources végétales de l'Afrique tropicale).* Wageningen University. Accessed 20 January 2023.

Oyen, L. P. A. (2008). Pterygota macrocarpa K. Schum. In D. Louppe, A. A. Oteng-Amoako, & M. Brink (Eds.), *PROTA (Plant Resources of Tropical Africa/Ressources végétales de l'Afrique tropicale).* Wageningen University. Accessed 20 January 2023.

Oyen, L. P. A. (2012). Celtis mildbraedii Engl. [Internet] Record from PROTA4U. In R. H. M. J. Lemmens, D. Louppe, & A. A. Oteng-Amoako (Eds.), *PROTA (Plant Resources of Tropical Africa/Ressources végétales de l'Afrique tropicale).* Wageningen University. Accessed 20 January 2023.

Padi, B. (1997). *Prospects for the control of cacao mealybugs.* Proceeding of the 1st International Cocoa Pests and Diseases Seminar. Accra, Ghana, November 6–10, 1995, pp. 249–263.

Rao, M. R., Palada, M. C., & Becker, B. N. (2004). Medicinal and aromatic plants in agroforestry systems. *Agroforestry Systems, 61,* 107–122. https://doi.org/10.1023/B:AGFO.0000028993.83007.4b

Rigal, C., Wagner, S., Phuong, M., Laurence, N., & Vaast, P. (2022). Shade-TreeAdvice methodology: Guiding tree-species selection using local knowledge. *People and Nature* (November 2021), 1–16. https://doi.org/10.1002/pan3.10374

Ruf, F. (2015). Diversification of cocoa farms in Côte d'Ivoire: Complementarity of and competition from rubber rent. *Economics and Ecology of Diversification, 15,* 1–340. https://doi.org/10.1007/978-94-017-7294-5

Ruf, F. O. (2011). The myth of complex cocoa agroforests: The case of Ghana. *Human Ecology, 39*(3), 373–388. https://doi.org/10.1007/s10745-011-9392-0

Schmelzer, G. H. (2006). Holarrhena floribunda (G.Don) T. Durand & Schinz. In G. H. Schmelzer & A. Gurib-Fakim (Eds.), *PROTA (Plant Resources of Tropical Africa/Ressources végétales de l'Afrique tropicale).* Wageningen University. Accessed 20 January 2023.

Schmelzer, G. H. (2008). Discoglypremna caloneura (Pax) Prain. In G. H. Schmelzer & A. Gurib-Fakim (Eds.), *PROTA (Plant Resources of Tropical Africa/Ressources végétales de l'Afrique tropicale).* Wageningen University. Accessed 20 January 2023.

Schmelzer, G. H. (2012). Daniellia ogea (Harms) Rolfe ex Holland. [Internet] Record from PROTA4U. In R. H. M. J. Lemmens, D. Louppe, & A. A. Oteng-Amoako (Eds.), *PROTA (Plant Resources of Tropical Africa/Ressources végétales de l'Afrique tropicale).* Wageningen University. Accessed 20 January 2023.

Schmidt, J. E., Duval, A., Isaac, M. E., & Hohmann, P. (2022). At the roots of chocolate: Understanding and optimizing the cacao root—Associated microbiome for ecosystem services. A review. *Agronomy for Sustainable Development, 42*(14), 1–19. https://doi.org/10.1007/s13593-021-00748-2

Sinmisola A., Oluwasesan B. M., & Chukwuemeka, A. P. (2019). Blighia sapida K.D. Koenig: A review on its phytochemistry, pharmacological and nutritional properties. *Journal of Ethnopharmacology, 235*, 446–459. https://doi.org/10.1016/j.jep.2019.01.017

Tcheghebe, T., Nyamen, L. D., Tatong, F. N., & Seukep, A. J. (2016). Ethnobotanical uses, phytochemical and pharmacological profiles, and toxicity of persea Americana mill: An overview. *Pharmacologyonline, 3*, 213–221.

Tchinda, A. T. (2008). Entandrophragma angolense (Welw.) C.DC. In D. Louppe, A. A. Oteng-Amoako, & M. Brink (Eds.), *PROTA (Plant Resources of Tropical Africa/Ressources végétales de l'Afrique tropicale).* Wageningen University. Accessed 20 January 2023.

Tchinda, A. T., & Tané, P. (2008). Amphimas pterocarpoides harms. In D. Louppe, A. A. Oteng-Amoako, & M. Brink (Eds.), *PROTA (Plant Resources of Tropical Africa/Ressources végétales de l'Afrique tropicale).* Wageningen University. Accessed 19 January 2023.

Tchoundjeu, Z., & Atangana, A. R. (2007). Ricinodendron heudelotii (Baill.) Pierre ex Heckel. In H. A. M. van der Vossen & G. S. Mkamilo (Eds.), *PROTA (Plant Resources of Tropical Africa/Ressources végétales de l'Afrique tropicale).* Wageningen University. Accessed 20 January 2023.

Todou, G., & Meikeu Kamdem, M. G. (2011). Musanga cecropioides R.Br. ex Tedlie. In R. H. M. J. Lemmens, D. Louppe, & A. A. Oteng-Amoako (Eds.), *PROTA (Plant Resources of Tropical Africa/Ressources végétales de l'Afrique tropicale).* Wageningen University. Accessed 20 January 2023.

Toirambe Bamoninga, B., & Ouattara, B. (2008). Morus mesozygia Stapf. In D. Louppe, A. A. Oteng-Amoako, & M. Brink (Eds.), *PROTA (Plant Resources of Tropical Africa/Ressources végétales de l'Afrique tropicale).* Wageningen University. Accessed 20 January 2023.

Tscharntke, T., Clough, Y., Bhagwat, S. A., Buchori, D., Faust, H., Hertel, D., Hölscher, D., Juhrbandt, J., Kessler, M., Perfecto, I., Scherber, C., Schroth, G., Veldkamp, E., & Wanger, T. C. (2011). Multifunctional shade-tree management in tropical agroforestry landscapes—A review. *Journal of Applied Ecology, 48*, 619–629. https://doi.org/10.1111/j.1365-2664.2010.01939.x

Twum-Ampofo, K. (2007). Albizia ferruginea (Guill. & Perr.) Benth. In D. Louppe, A. A. Oteng-Amoako, & M. Brink (Eds.), *PROTA (Plant Resources of Tropical Africa/Ressources végétales de l'Afrique tropicale).* Wageningen University. Accessed 19 January 2023.

Vaast, P., Harmand, J. M., Rapidel, B., Jagoret, P., & Deheuvels O. (2016). Coffee and cocoa production in agroforestry—A climate-smart agriculture model. In T. Emmanuel (Ed.), M, David & C. Paul (Trans.), *Climate change and agriculture worldwide* (pp. 197–208). Springer.

van Vliet, J. A., & Giller, K. E. (2017). Mineral nutrition of cocoa: A review. *Advances in Agronomy, 141*, 185–270.

van Vliet, J. A., Slingerland, M., & Giller, K. E. (2015, July). Mineral nutrition of cocoa. *Advances in Agronomy.* https://doi.org/10.1016/bs.agron.2016.10.017

Yerou, K. O., Ibri, K., Bouhadi, D., Hariri, A., Meddah, B., & Touil, A. T. (2017). The use of orange (*Citrus sinensis*) peel as antimicrobial and antioxidant agents. *Journal of Fundamental and Applied Sciences, 9*(3). https://doi.org/10.4314/jfas.v9i3.7

Young, A. (1990). Agroforestry for soil conservation. In *Soil erosion and conservation*. BPCC Wheatons Ltd. https://doi.org/10.1016/0308-521x(91)90121-p

Zimudzi, C., & Cardon, D. (2005). Morinda lucida Benth. In P. C. M. Jansen & D. Cardon (Eds.), *PROTA (Plant Resources of Tropical Africa/Ressources végétales de l'Afrique tropicale)*. Wageningen University. Accessed 20 January 2023.

Open Access This chapter is licensed under the terms of the Creative Commons Attribution 4.0 International License (http://creativecommons.org/licenses/by/4.0/), which permits use, sharing, adaptation, distribution and reproduction in any medium or format, as long as you give appropriate credit to the original author(s) and the source, provide a link to the Creative Commons license and indicate if changes were made.

The images or other third party material in this chapter are included in the chapter's Creative Commons license, unless indicated otherwise in a credit line to the material. If material is not included in the chapter's Creative Commons license and your intended use is not permitted by statutory regulation or exceeds the permitted use, you will need to obtain permission directly from the copyright holder.

CHAPTER 4

Social Challenges and Opportunities in Agroforestry: Cocoa Farmers' Perspectives

Aske Skovmand Bosselmann, *Sylvester Afram Boadi*, *Mette Fog Olwig*, *and Richard Asare*

Abstract Agroforestry practices in cocoa cultivation have historical roots going back to the Mayan sacred groves in Mesoamerica. Today, agroforestry cocoa, i.e., the integration of shade trees, plants and crops in cocoa systems, is promoted as a climate smart practice by public and private institutions. Shaded cocoa can sustain or even increase cocoa yields and the agroforestry systems may provide additional output for household consumption and sale as well as improve the microclimate and soil conditions on the farm. Despite these promising features, cocoa agroforestry systems are far from the norm in producing countries like Ghana. Based on discussions with groups of farmers across the Ghanaian cocoa belt, this chapter shows that while farmers are well aware of the positive aspects of

A. S. Bosselmann (✉) · S. A. Boadi
Department of Food and Resource Economics, University of Copenhagen, Frederiksberg, Denmark
e-mail: ab@ifro.ku.dk

S. A. Boadi
e-mail: saboadi@csir.org.gh

© The Author(s) 2024
M. F. Olwig et al. (eds.), *Agroforestry as Climate Change Adaptation*,
https://doi.org/10.1007/978-3-031-45635-0_4

shaded cocoa systems, traditional cocoa practices, village chiefs' command of local land uses, land and tree tenure systems, alternative land uses and inability to access inputs and extension services limit the adoption and constrain the management of shade trees. As still more policies are developed to improve the Ghanaian cocoa sector, policymakers must consider these often overlooked social and institutional factors that prevent cocoa farmers from engaging in longer-term agroforestry practices and thereby benefiting from the opportunities they present.

Keywords Land and tree tenure rights · Multi-institutional complex · Non-timber forest products · Smallholder perspectives · Mining activities · Socio-cultural and gender dimensions

4.1   INTRODUCTION

Cocoa agroforestry systems are described as climate smart practices because of their potential ability to mitigate and adapt to climate change, while ensuring diverse farm outputs (Vaast et al., 2015). However, the cultivation of cocoa in intercropping or shaded systems is not a new practice developed in the face of climate change. Quite the contrary, cocoa has been cultivated under shade trees since the domestication of the *Theobroma cacao* tree in pre-Hispanic South and Central America (Gómez-Pompa et al., 1990). As a highly valuable crop used for religious

S. A. Boadi
Department of Geography and Resource Development, University of Ghana, Accra, Ghana

CSIR-Water Research Institute, Accra, Ghana

M. F. Olwig
Department of Social Sciences and Business, Roskilde University, Roskilde, Denmark
e-mail: mettefo@ruc.dk

R. Asare
International Institute of Tropical Agriculture (IITA), Accra, Ghana
e-mail: r.asare@cgiar.org

ceremonies, as food and as currency, the Mayan cultivated cocoa trees in sacred groves, either in agroforestry systems or in sinkholes, where steep slopes and high soil humidity provided an adequate yet geographically very limited microclimate (ibid.). Cocoa continues to this day to be part of religious practices and has ceremonial value for modern-day Maya groups. The shaded cocoa habitats have even been described as limiting deforestation and forest degradation due to the sacred character of the cocoa trees (Kufer et al., 2006; Steinberg, 2002). Cocoa cultivation in shaded systems, often intercropped with other food crops, has persisted in various forms as a central practice in traditional cocoa farming from Latin America to West Africa and Indonesia (Oladokun, 1990; Rice & Greenberg, 2000; Schulz et al., 1994). Today, shaded cocoa cultivation may take many forms, ranging from cultivation in the limited shade provided by a single tree species, often timber trees, to rustic shade systems, where cocoa is found under remnant forest trees, to a more actively managed fully fledged cocoa agroforestry system with several strata, each consisting of multiple trees with diverse purposes (see typology in Orozco-Aguilar et al., 2021). However, intensive cocoa farming with high performing cocoa varieties in lightly shaded or full-sun conditions is currently the rule rather than the exception. As a result, cocoa is more often mentioned as a deforestation driver and less as a harbourer of biological diversity (Franzen & Borgerhoff Mulder, 2007; Ordway et al., 2017; Ruf & Schroth, 2004), especially in Côte d'Ivoire and Ghana (Kalischek et al., 2022).

Cocoa agroforestry has been highlighted for its ability to increase the total economic output from cocoa and shade trees, take advantage of cost complementarities between cocoa and other products on the same plot, and reduce input dependencies in low-input systems managed by smallholder farmers with limited access to fertilizers and pesticides (Herzog, 1994; Ofori-Bah & Asafu-Adjaye, 2011). For example, shade trees may provide nutrients and humidity to the soil through branch pruning and litter decomposition, and provide farmers with tree products, such as edible plant parts, firewood, timber, fibers and fodder, both for subsistence use and for sale (Bos et al., 2007; Graefe et al., 2017; Kaba et al., 2020). Products from shade trees may thus have a role as an income gap filler, while timber trees may function as a safety net during times of low income. Yet, while cocoa plantations are often established under shade through intercropping to shield the young plants, mature plantations in West Africa often become monocrop systems to avoid cocoa trees competing with shade trees for water, nutrients, space and light. However, while competition between shade trees and cocoa has been documented, e.g., for soil water in situations with prolonged droughts (Abdulai et al., 2018), limited effects or even a positive effect on cocoa yields have been found in systems with low to moderate levels of shade (Abou Rajab et al., 2016; Asare et al., 2019; Nunoo & Owusu, 2017). The positive role of shade trees on economic output is further augmented when the prolongation of the main productive phase of cocoa trees, due to the presence of shade, is considered (Asare et al., 2019). While full-sun cocoa systems have become widespread in the search for higher yields, research is increasingly finding that cocoa agroforestry systems, when appropriately implemented and managed, may outperform full-sun systems on economic as well as environmental parameters (see also Chapters 3 and 5 in this volume).

With the advent of human-induced climate change, agroforestry is increasingly being highlighted as a climate smart practice, especially in perennial cropping systems such as cocoa cultivation. Climate smart practices entail adaptation to long-term climate change and erratic weather events, climate mitigation by reducing the emissions of greenhouse gasses and possibly sequestering gasses from the atmosphere, and sustainably increasing the productivity of the agricultural system (FAO, 2009). In West Africa, where the main share of the global cocoa production takes place, the effects of climate change are exerting pressure on cocoa farmers to change crops or adopt climate smart practices to adapt to higher temperatures and change in precipitation patterns. In their recommendation across different agro-ecological zones in Ghana, Bunn et al. (2019) emphasize the use of shade trees to adapt to climate change.

Cocoa agroforestry systems are being promoted in voluntary certification schemes as well as in corporate programs for responsible cocoa production and sourcing in which almost all major cocoa buying companies are engaged (Carodenuto & Buluran, 2021; Thorlakson, 2018). These corporate initiatives will very likely gain further traction as new public regulations and directives for the main market for cocoa and chocolate, the EU, are expected to push the agenda for deforestation-free cocoa without climate emissions. An EU deforestation regulation and the Corporate Sustainability Due Diligence Directive (CSDDD) will hinder any trade to the EU of cocoa unless the trading company can document that the cocoa is not associated with deforestation, is legally produced and does not have any adverse climate impacts. While neither has been implemented yet, cocoa buying companies are setting up programs for deforestation-free cocoa that also promote dissemination of shade tree seedlings (Nasser et al., 2020).

Having in mind the agri-ecological benefits of shaded cocoa production, the long-term benefits of cocoa agroforestry to the farming household, as well as its promotion as a climate smart practice and part of a sustainable business model, it seems surprising that agroforestry is not the dominant way of producing cocoa. Kaba et al. (2020) relate the low adoption of agroforestry to a mismatch in farmers' and researchers' understanding and perception of shade tree integration in cocoa farming. Farmers generally possess knowledge of the positive and adverse effects and outcomes of intercropping trees and cocoa, as shown in several studies (e.g., Awuah & Kyereh, 2019; Graefe et al., 2017; Smith Dumont et al., 2014). There are seemingly other factors at play that keep farmers from returning to the old ways of the Mayan shaded cocoa groves and that influence farmers' decision and ability to plant and care for trees in their cocoa plantation. Based on discussions with cocoa farming communities in Ghana and interviews with key informants in the Ghanaian cocoa sector, this chapter explores and discusses the social challenges as well as opportunities linked to agroforestry from the perspective of the cocoa farmers. The following section provides further background on farmers' valuation of trees in cocoa cultivation and the obstacles that may limit farmers' ability and willingness to plant trees, mainly based on studies from Western Africa. This section is followed by a discussion of the experiences of Ghanaian cocoa farmers, who are struggling along several fronts concerning the integration of trees in cocoa farming. Finally, possible pathways for facilitating the integration of shade trees in cocoa farms are presented.

4.2   Background

Cocoa cultivation in West Africa goes back to the 1880s and has long been one of the main income-generating activities that support the livelihoods of millions of farmers in Côte d'Ivoire and Ghana. In 2019, according to UN trade data,[1] cocoa provided around USD 2.7 bn. in export earnings to Ghana through exports of cocoa beans and other cocoa

---

[1] https://comtrade.un.org/data, Trade codes HS 1801–1806.

products. The parastatal Ghana Cocoa Board (COCOBOD) obtained revenues of around USD 1.3 bn., which among other things covered large-scale service provision programs to farmers. Roughly, USD 900 million was paid to farmers, while the remaining earnings were captured by traders and grinders with domestic operations in Ghana. These figures only tell part of the story of the importance of cocoa to rural communities in Ghana. Other types of economic and social values exist in relation to the cocoa cultivation systems and the intercropping of trees.

4.2.1   Farmers' Cocoa Agroforestry Economy

There are far more studies of the economic value of the cocoa crop than of the value of shade trees in cocoa agroforestry systems. Nevertheless, several studies have highlighted the potentially extensive values of shade tree products and ecosystem services that farmers may obtain when managing cocoa farms for more than just cocoa. In cocoa agroforestry systems in Southern Cameroon, Gockowski et al. (2010) recorded 286 different plant species that farmers used for food, medicine, timber, packaging materials and other non-timber forest products. The non-cocoa products generated 217 USD/ha in one area, compared to 425 USD/ha from cocoa, and across all regions, trees and plants generated 25% of total farm income, mainly driven by sales of palm oil, timber and fruits. While Gockowski et al. focused on marketed products, Cerda et al. (2014) also included the value of the households' own consumption of non-cocoa products in their research on cocoa farmers in Central America. The authors found that the economic benefits to the households of bananas, fruit trees and timber in the cocoa plots equaled or exceeded the family income from cocoa sales. Obeng et al. (2020) went a step further and used contingent valuation methods to assess Ghanaian farmers' willingness to pay for tree integration on their cocoa farm in order to obtain non-marketed ecosystem services, such as erosion control, temperature regulation and water resources protection. They estimated the value of bundled ecosystem services to be USD 164 per ha per year, corresponding to 8.2% of the farmers' cocoa income.

In the study by Obeng et al. (2020), farmers' willingness to pay for tree integration was significantly influenced by their positive attitude toward forests in general. Farmers emphasized the existence value of tropical forest more so than its current use value as their motivation for off-farm forest protection. However, more tangible values were prioritized for on-farm tree integration. The shading of cocoa trees, access to timber and (nature) medicine and environmental benefits such as providing a habitat for pollinators were among the main reasons behind farmers' valuation of tree integration (ibid.).

The benefits of shade tree products may come at the cost of reduced cocoa yields, though the results from studies of combinations of cocoa and different shade trees vary. Koko et al. (2013) find that intercropping of fruit trees in Côte d'Ivoire reduced yields per cocoa tree and per ha. They also cite a number of earlier studies from Latin America that showed similar results and argue that excessive light inception reduces flowering and thus yields. Results from more recent studies paint a different picture. Abou Rajab et al. (2016) find no negative effect on cocoa yields between monocultures and multi-shade systems in Indonesia, while Asare et al. (2019) document a doubling of cocoa yields in Ghana when changing from no shade to 30% canopy cover. Equally important, Nunoo and Owusu (2017) find that shade increases the length of the mature cocoa producing phase based on data from Ghana, thus prolonging the economically productive phase of the rotation length. Despite lower yields under shade, Koko et al. (2013) obtain much higher yields in their trial experiment in their shaded plots than the average productivity in West Africa. This points to another important factor: access to inputs, farmer skills and management priorities highly affect cocoa yields, with or without shade trees. Whereas the cocoa plots in Koko et al. (2013) received optimal levels of inputs to maximize yields, we found that many small-scale farmers do not have access to or cannot afford fertilizers and agrochemical inputs. They therefore tend to aim for lower but stable outputs based on more "nature-based solutions"—to borrow a term from climate smart agricultural programs—which include shade trees for weed suppression, soil fertilization and moisture, and food.

Important factors when cocoa farmers make overall farm management choices include trade-offs between different crops, access to inputs required for different systems, and the value of ecosystem services from shade trees. From this perspective, the intercropping of shade trees in cocoa plots represents an opportunity for added value, especially when farmers do not have adequate access to inputs or for other reasons manage their cocoa plot extensively. This is exemplified in the study by Bentley et al. (2004), who found that more diverse agroforestry was practiced mainly by farmers with low-input management regimes in Ecuador. The economy of cocoa agroforestry may thus be particularly advantageous for smallholders.

4.2.2   The Socio-cultural Context of Cocoa Agroforestry Systems

At the place of origin of the cocoa tree in South America and Mesoamerica, many farmers attach ceremonial and social values to their cocoa plots. From Mexico and Guatemala to the Ecuadorian Amazon region, cocoa trees and associated forest trees are regarded not only as a productive system, but also as a social-ecological system with deep-rooted cultural values (Coq-Huelva et al., 2017; Kufer et al., 2006). This is a result of cocoa cultivation having evolved along with other societal developments over several thousand years in South America (Zarrillo et al., 2018). In contrast, the cocoa tree was introduced as a cash crop in Ghana in relatively recent times, just 150 years ago. The ritual and ceremonial values attached to cocoa at its natural origin did not accompany the beans on the Portuguese ships that first brought the crop to West Africa (Ryan, 2011). However, the economic value and importance of the crop in Ghana, which has surpassed the crop's economic importance on the American continent, have influenced Ghanaian culture beyond the economic aspects.

"Ghana is cocoa, cocoa is Ghana" is a common saying in the world's second largest cocoa producing country, not only because of the thousands of cocoa farming households and the nearly one million people working in the cocoa plantations. For better or worse—"worse" referring to a colonial history and coercive use of labor, including child labor, that still taint the cocoa sector today—cocoa has shaped Ghanaian society since its introduction. Tetteh Quarshie, the blacksmith most often accredited with the first introduction of cocoa to Ghana, is regarded as a national hero and is a figure that continues to be present in the cocoa sector and in society; several streets, a highway and a hospital bear his name, as does one of the traditional cocoa varieties.

Less obvious is the connection of agroforestry practices in cocoa farming to local cultures and cultural values. Mr. Quarshie established the first cocoa farm in Ghana, intercropping a variety of food crops. This practice continues today, as many farmers rely on food crop production in and around their cocoa farms, for sale and consumption. Intercropping helps to shield the young cocoa plants and to gap-fill farm outputs in the first years of the plantation, but is then often abandoned for a sole focus on cocoa yields. Cocoa has been one of the main drivers of deforestation in Ghana, and still is today (Acheampong et al., 2019), but some farmers do retain forest trees as they move into new areas or plant trees for a variety of reasons. The choice of trees is mainly related to the use and economic value of the trees (Anglaaere et al., 2011), but what is useful and of economic value depends at least partly on the socio-economic context and the local knowledge of trees (see Appendix in Chapter 3 in this volume). Lack of education providing awareness of tree benefits has been identified as a key factor explaining why Ghanaian farmers remove shade trees (Kaba et al., 2020). Specific knowledge of shade tree—cocoa compatibility in Southern Ghana and needs for income diversification in the Northern cocoa areas of Ghana have been found to be important for shade tree selection (Graefe et al., 2017). Gender aspects also play a role, as women more often select shade trees for household consumption purposes than men, but at the same time, they may be constrained in terms of shade tree management due to intra-household power dynamics, lack of land possession and access to hired labor, as found by Jamal et al. (2021).

In a very different part of the world, East Papua New Guinea, where cocoa is also a relatively recent crop, the socio-cultural context plays a different role in cocoa cultivation. Low-input cocoa systems with diverse intercropping are favored by the local traditional "way of life" and moral values and are seen as providing status and identity (Curry et al., 2015). While the social obligations to share surplus generate community-wide benefits, they also create a socio-cultural context that limits farmers' ability to invest in and build on savings from cocoa cultivation. The same social constraints are not found in Ghana, where it seems the main cultural value of agroforestry systems is related to economic outputs and the viability of the productive agro-ecological system, even though the socio-cultural context influences the importance given to different types of shade trees, such as multipurpose trees for materials, medicines and other sources of environmental income.

4.2.3   The Multi-institutional Complex of Shade Tree Systems

The care for trees and the right to enjoy the benefits of harvesting the trees is no simple matter in Ghana, legally speaking. National land tenure policies, tree use permits, traditional land rights vested in the chief, and a mix of matrilineal and patrilineal inheritance systems that tend to weaken individual land rights (Quisumbing et al., 2001), further complicated by a long history of domestic migration, come together in a hot pot of rights, obligations and opportunities that influence cocoa farmers' willingness and ability to invest in and manage shade trees.

According to Mayers and Ashie Kotey (1996), land tenure is influenced by tradition, politics and postcolonial policies. Traditional landholding authorities, most often the chief, may be a paramount, divisional or sub-stool or a combination of these, depending on the mode of acquisition of the land and history of the people. Chiefs may hold absolute title to land on behalf of the people, who in turn have usufruct rights and can appropriate a portion of the land for permanent development (Mayers & Ashie Kotey, 1996). Such land, for most practical purposes, belongs to the community member with usufruct rights whose interests should be secure, inheritable and generally alienable (Spichiger & Stacey, 2014). Migrants acquire land by outright purchase, or more commonly by leasing under customary law. Traditional authorities may also grant tenancies on *abunu* terms for cash crops, where e.g., cocoa land is shared between landowner and tenant once the cocoa crop is mature, or on *abusa* terms for food crops where the food production, not the land, is shared. Many poorer migrant farmers are in *abusa* arrangements, which are generally insecure and therefore create little incentive to plant and nurture trees (Mayers & Ashie Kotey, 1996).

While the farmers' right to maintain trees on cocoa farms and to have their farms protected from timber concessionaires has been in place since 1979, only from 1995 was it possible for farmers to receive compensation for crop damages incurred when timber was harvested on their fields. All revenues from the timber, however, were to be divided between public authorities and the traditional authorities (Amanor, 1996). The legal basis for farmers to refuse the felling of timber on their farmland or negotiate a price for each tree to be felled by a concessionaire was finally established with the Timber Resources Management (Amendment) Regulations in 2003. However, the full power of landholders to plant, maintain, harvest and sell timber from their own land is still a wishful scenario for most cocoa farmers. Today, rights to timber on farmland may still be afforded to concessionaires despite farmers' rights to refuse, and farmers have to navigate a bureaucratic registration system to register trees and gain user rights to individual trees on their farm (Gaither et al., 2019; Hirons et al., 2018). For these reasons, farmers with timber trees on their farmland will often gain greater benefits from engaging with the informal wood sector, rather than trying to stay within the formal legal system (Hirons et al., 2018).

Within this complex institutional framework for land and tree rights corporate sustainability programs run by international cocoa buyers and chocolate companies are disseminating shade tree seedlings to farmers across Ghana. Tree seedlings are part of a bundle of services farmers receive, either at no cost or paid back via a share of the harvest, as part of the cocoa-industry's efforts to build capacity in the Ghanaian cocoa farming communities vis-a-vis yield improvement, climate resilience, sustainability, regenerative practices or other objectives often found in industry financed, farmer facing programs (Boadi et al., 2022; Carodenuto & Buluran, 2021; Nasser et al., 2020). In this context, women are also marginalized in terms of access to shade trees and training, because programs most often interact with male landowners and focus on technical solutions (e.g., number of tree seedlings distributed) rather than gender differentiated solutions (e.g., female farmers' selection of tree species) (Friedman et al., 2018). It is within this multi-institutional context that farmers must navigate when they consider the short and long-term costs and benefits of planting, maintaining and harvesting different types of shade trees for different kinds of purposes, besides "simply" shading the cocoa trees.

4.3   Talking About Shade Tree Management with Ghanaian Cocoa Farmers

In order to improve our understanding of Ghanaian cocoa farmers' perceptions of shade trees and their associated values in cocoa agroforestry systems, along with the socio-cultural context defining the challenges and opportunities related to cocoa agroforestry, we organized a number of focus group discussions with cocoa farmers across the cocoa zones in Ghana. In total, 20 focus group discussions were carried out with female farmers, male farmers and in mixed groups over a period spanning 2018 and 2019, covering 12 villages in the districts of Asutifi South, Offinso North, Amansie West, Atwima Nwabiagya, Sefwi Wiawso and Wassa Amenfi. The participants were selected from a pool of over 400 farmers, who had participated in a farmer survey on cocoa production practices and shade tree management (see Chapter 5 in this volume for details and map of study areas). They represented both migrant farmers and farmers native to the cocoa communities, as well as farmers with private landholdings and farmers in shared land arrangements. The expected differences in use of different shade tree species, given possible differences in knowledge about trees and access, were not clear in the quantitative survey data. While native farmers as a group had more species represented, they also accounted for most of the cocoa plots registered in the survey, and many species were only found on single cocoa plots. Both groups of farmers favored a handful of timber and fruit tree species. Several species of trees with medicinal uses were also mentioned, but were found on fewer plots.

Sitting outdoors in shaded environments under community trees, we began our discussions by listening to farmers' history of their community or their stories of migration to existing cocoa areas often several generations ago. Then we centered on the use of shade trees in cocoa cultivation, the benefits and disadvantages of specific shade trees or, more generally, of shade in cocoa farming, and the challenges and opportunities experienced by farmers who either wanted to manage or already managed tree-cocoa intercropping systems. Furthermore, in each group discussion, the issues of climate change and possible coping strategies, land tenure and the future of cocoa farming were addressed, as these issues are often studied in the context of agroforestry. While the topics were predefined, the discussions were allowed to make detours to related topics, often as a result of disagreements among participants, such as farmers' perceptions of the practice of small-scale gold mining, also referred to as *galamsey*, and sand mining. The farmers saw mining activities as either new avenues for income and livelihood improvements, or as detrimental to future economic activities and agricultural-based living, but all agreed that mining conflicts with cocoa and tree management.

Over the course of the 20 group discussions, we heard the views of 70 male and 53 female cocoa farmers. Their views were supplemented with four in-depth interviews with cocoa buying agents and lead farmers in some of the communities. The qualitative data was transcribed and analyzed for commonalities across discussions and locations that would improve our understanding of the challenges and opportunities of shaded cocoa as experienced by the farmers.

4.4 Recent Perspectives
from Ghanaian Cocoa Farmers

The following represents the voices, perceptions and experiences that were common to most, if not all, groups of farmers participating in the focus group discussions, complemented by references to previous studies in the Ghanaian cocoa belt.

4.4.1 Common Practices, Changing Practices

> Since we took over from our grandparents, we have maintained their farming practices. We cut down all the trees, after which we burn the weeds on the land. (Discussion participant, Esaase Community)

Whether farmers were native or had migrated to their current community, the majority of farmers described using the traditional slash-and-burn approach to establish their cocoa plots and intercrops mainly to shade the young cocoa plants; practices that the farmers have learned from observing and participating in the cocoa activities of parents, other family members and community members. Even recent plot establishment had required clearing and burning forest areas or fallows. However, during all discussions, participants acknowledged the protective effect of shade trees and described how they bring about a better and cooler climate and more humidity and help cocoa to survive during warm periods. Most farmers indeed described having smaller shaded areas or a few shade trees dispersed on their cocoa plots. Farmers without shade on their own farms, described visits to neighboring farmers, whose shaded cocoa trees were performing better during warm periods, while their own had "their tops burned off" as one farmer described it. Even farmers with negative perceptions of shade effect on cocoa yield and presence of pests recognized the positive role of shade trees on the microclimate. This recognition was seemingly related to farmers' account of recent experiences of changes in rainfall patterns, longer droughts, higher temperatures and intense sunshine. Many farmers agreed that at least some level of shade was necessary throughout the cocoa trees' lifetime, and some described how they had recently introduced the first or additional shade trees in mature cocoa plots, on a needs basis, referring to the specific service of shading—though often combined with reference to timber.

There was, however, also consensus on keeping shade trees limited in number, so as not to decrease the cocoa yields.

Farmers' choice of shade tree species and their management were mainly informed by agricultural advisory and tree planting schemes, but also based on the farmers' own experiences. Avocado, orange and mango were the most common fruit trees due to their economic benefits, while a number of valuable timber trees were favored, partly for economic reasons and partly due to "tall and broad trees" being good characteristics of shade trees according to advice from extension officers. Contrary to this, there was less consensus when it came to selecting tree species believed to have medicinal properties, which is related to differences in the specific knowledge of individual farmers. Similarly, information on trees with detrimental effects on cocoa farming was to a larger extent based on farmers' own experiences with how certain trees limit the growth of the cocoa trees or attract pests, even trees recommended by extension officers.

Research is emerging on the impact of different shade tree species and shade tree species diversity in cocoa agroforestry systems (Asare et al., 2019; Asitoakor et al., 2022; Graefe et al., 2017; Kaba et al., 2020. See also Chapters 3 and 5 in this volume). We found that some extension services that support better shade tree selection and shade tree seedlings are being offered to a limited group of farmers. Indeed, we found that these farmers were most able to implement cocoa agroforestry systems successfully. Yet, even though NGOs, cocoa-industry initiatives and the state all provide extension services, most farmers only receive limited training and are only given seedlings from a very small selection of shade tree species. The farmers' discussions did indicate smaller changes in their perceptions and management of shade trees, but changing policies and the lack of consistency of the agri-advisory services offered to farmers have created some distrust among farmers of advice from institutions offering such services, including advice on shade tree integration. Farmers thus reported conflicting recommendations from extension services and described how they were first recommended to eradicate shade trees as they were not good for cocoa, but later on, the same extension services came back to recommend tree integration. There is a need for consistency in terms of agri-advisory regarding tree integration in cocoa farms, but also better communication concerning desirable/undesirable shade trees, the contextual nature of what constitutes an optimal number of shade trees to manage on cocoa farms and the importance of shade tree species diversity.

4.4.2   Environmental Income and Shade Tree Products

> These trees have been helping me a lot. Quite recently, I went to harvest some of these trees when I was in financial hardship. It is good and beneficial to nurture such trees in the farm. (Focus group participant, Nerebehi, Ghana, talking about timber trees in cocoa)

Cocoa farmers with a more diverse shade tree composition and intercropping were able to reap monetary benefits additional to the sale of cocoa (Chapter 5 in this volume). The farmers' discussions revealed three broad categories of benefits derived from shade trees and agroforestry: (i) tree products, such as fruit, timber and fuel wood, (ii) benefits to the cocoa system itself, such as improved water retention and protection from high temperatures and direct sunlight and (iii) harvesting of mushrooms and snails from the shaded environment on and adjacent to the cocoa farms. The latter used to be important for many farmers, both for sale and consumption, but most farmers were now only reminiscing about the time when there was an abundance of mushrooms and snails on or near their farms. Very few farmers reported currently collecting mushrooms and snails blaming general deforestation, use of pesticides in cocoa and other agriculture, and bushfires to be the culprits of the disappearance of this environmental income, confirming a trend documented in 2008 by Ahenkan and Boon (2011). Farmers acknowledged how planting of shade trees within the limits of their own farms was not sufficient to provide the habitat for snails and mushrooms, as not only a shaded environment is needed but also decaying wood and thick undergrowth on larger areas. Some farmers jested of having to buy cultivated mushrooms and snails, others talked of missing a piece in their diversified livelihoods.

As a result of changing landscapes, trees on the farm have become the source of environmental income. The multiple tree products mentioned by farmers included firewood, various fruits, leafy vegetables and food ingredients, tree parts or sap with medicinal properties, along with various construction materials that are mainly used on their own farms. The access to tree seedlings was a recurrent subject, as some trees—those most favored by farmers due to multiple products afforded by a single tree—were difficult to regenerate naturally and farmers therefore were dependent on buying seedlings. When selecting shade tree species, farmers considered not just the potential added benefits, but also the problems that could arise when including specific species in agroforestry, including competition for water and nutrients and an increase in pests and diseases (see also Chapter 3). Avocado, a good source of fruits for sale and consumption, was known to attract mistletoe that would also negatively affect certain varieties of cocoa, while oil palms were favored by some as a food ingredient, but reportedly harbored squirrels and "destroyed cocoa trees." Furthermore, some trees with medicinal value were not thought to be compatible with cocoa farming. Farmers must therefore carefully select trees for different purposes.

Shade trees of timber species, cared for with the intention of harvesting poles and beams for constructing and roofing houses, were among the most contentious issues discussed by the farmers. Timber trees may support families during hardship if sold on the market, as exemplified in the quote above, but farmers were well aware of the complex set of rules that surrounds timber trees and restricts the use and sale of timber, even of trees planted and cared for on private farmland. Some farmers even resorted to removing valuable trees before maturity to avoid trouble and, in no small part, out of spite of the Forestry authorities. Doing that, they also forgo what may be a substantial value from the cocoa systems, as documented by Nunoo and Owusu (2017) and Obiri et al. (2007) among Ghanaian cocoa farmers.

4.4.3   Gold and Sand Mining—Competing and Destructive Land Uses

> They are profiting from the mining operations, but we are dying. What are we going to do as the government has given the mining companies the permit to mine in the mountains which is the source of all our waterbodies, and as the activities of these mining companies is resulting in the breaking apart of the mountain and the cutting down of the trees? (Focus group participant, Jeninso, Ghana)

The quote introducing this section represents the situation in five of the 12 communities, where the focus groups discussions took place. Along with sand winning, *galamsey* activities, or small-scale mining,[2] were seen by especially the older cocoa farmers to be among the largest threats

---

[2] Galamsey is derived from the phrase "gather them and sell," and is used to describe illegal, small-scale mining activities, mainly for gold.

to not only cocoa farming but also agricultural activities in general. Farmers described the activities mainly as illegal activities, often accepted or even facilitated by the local chief, and carried out on a small scale by people not from the local community or on larger tracts of land by mining companies. Farmers described *galamsey* and sand winning activities as leading to destroyed roads and footpaths, complicating access to farms, increasing the occurrence of forest fires, impacting water bodies, uprooting cocoa trees and removing soil cover, thereby leaving farmland unproductive. The loss of land had incentivized some farmers to look for forested areas to establish new cocoa farms, indicating a push factor from mining activities leading to cocoa-related deforestation. The farmers also associated the mining activities with a lack of labor for agricultural activities, as day wages cannot compete with the possible earnings of mining activities. Older farmers told of conflicting views; they discouraged their children from pursuing *galamsey* activities, but also acknowledged the hardship and risky livelihood related to cocoa farming in a context of other and faster economic opportunities. This argument was also voiced by younger farmers participating in the discussions. Nonetheless, they did not consider engaging in mining activities.

"If you find it, you own it" read the sign of a large mining company that flanked the entrance to a community where one of the focus group discussions took place. The advert seemed to have worked; along the local water bodies and in-between cocoa farms, pits and mounds of gravel from *galamsey* activities characterized the landscape. This was not a lone incident. Across Ghana, an estimated 300,000–500,000 small-scale, unlicensed miners are supporting an industry worth millions of dollars annually, often acquiring farmland from cash strained farmers (Siaw et al., 2023). Small-scale mining, when regulated, is seen as an economic activity that can help to alleviate poverty in rural areas of Ghana (Okuh & Hilson, 2011), but *galamsey* may also be seen as the antithesis to cocoa agroforestry farming. *Galamsey* favors short-term benefits at the cost of arable land, and cocoa farming is a long-term strategy for climate smart agriculture. For both, a facilitating regulatory and policy environment is needed to promote socio-economic development (Ofosu & Sarpong, 2022), but for cocoa agroforestry practices not to lose out to mining activities in overlapping areas, strong long-term incentives are needed from both public and private actors. These include secure land and tree rights as well as relevant pricing mechanisms for cocoa from shaded systems.

4.4.4   Rights or No Rights to Land and Trees

There is a law, which forbids a farmer from harvesting the trees which he has planted on his farm and that have matured; there is a law which calls for the arrest of any farmer who commercializes tree harvesting. (Focus group participant, Mehame, Ghana)

Without being able to name the many policies and laws governing land tenure and tree rights, many farmers did clearly communicate the trouble of living with the uncertainty and complexity of rules and powers affecting access to land and trees. Some farmers, mainly natives to the communities, expressed having secure land rights and described how even if the local chief were to invite mining companies to mine their plots, or timber contractors to harvest the trees, the farmer would still be the one benefitting. Other farmers held deep, negative perceptions of the chiefs and shared experiences of chiefs who allocated the farmers' cocoa plots to sand winning and *galamsey*, or the timber trees to outside chainsaw operators without consulting them. Farmers described returning to their cocoa plots, only to find food crops and cocoa trees removed along with the topsoil, leading to the loss of livelihoods. In other narratives, the cocoa plots were allocated to urban extensions. Some farmers accepted this. Even after several generations of staying in the same community, farmers explained that they owe their land endowments to the village chief and therefore accept the chief's decision-making power over land allocations and use. Others were more frank in their assessment of the chiefs' "destroying our lands" for their own gain, but also described how little could be done about it and the fear of arrest if complaints were to be made.

Chiefs and elders of the communities were also mentioned as being involved in matters of timber trees on cocoa plots, but more often farmers referred to regulations implemented by officers from the Forestry Commission. A few farmers asserted full rights over trees grown and harvested on their farms, even when in sharecropping arrangements, and some described how tree materials could be used for their own houses, such as roofing, sometimes after consulting the chief and/or landowner. Many more were acutely aware of the limitations of harvesting trees, whether for sale or own use, and acknowledged the need to register individual trees and secure permits at the local Forestry Commission office in order to secure rights to the trees on their cocoa plots. The individual tree registration at the Forestry Commission, much akin to how the Land Commission should register land allocations in customary lands (Spichiger & Stacey, 2014), is the main approach adopted by public authorities to address tree tenure issues and create clarity of ownership and rights to usage. The farmers, however, often saw it as a way for public authorities to collect payment as a fee is paid for registering trees, and instead of a solution, farmers view tree registration as yet another source of tree rights disputes. Other farmers, especially those who had acquired their plots via the *abunu* sharecropping arrangement, referred to agreements that revert the land, and any timber/shade trees planted on it, to the landowner when the standing cocoa trees come to the end of the rotation cycle. Such agreements hindered not only shade tree integration, but also cocoa farm rehabilitation and renovation.

Some farmers had received tree seedlings from agricultural extension officers and, along with them, the rights to the tree. With the same aim, cocoa buying companies are disseminating tree seedlings to cocoa farmers to promote agroforestry practices in their supply chains, but even for large multinationals, the administrative burden of documenting and registering trees has led to projects giving up on tree registration, relying instead on traditional rules (O'Sullivan et al., 2018).

Insecure land and tree tenure regimes impede farmers' willingness to make long-term investments in their cocoa plots, including the planting and tending of timber trees in cocoa agroforestry systems. Indeed, for some farmers, the insecurity of tree ownership was seen as an incentive to remove shade trees.

4.4.5   Policy Implications—Private and Public

Secure long-term rights to land and trees are necessary for farmers to carry the long-term investment in cocoa agroforestry systems. While the egalitarian objectives of the formal state laws and traditional land authorities do exist on paper, the missing checks and balances that should exist between the different layers of customary land governance and administration units, and thus the missing accountability of chiefs, result in uncertainties and land conflicts (Spichiger & Stacey, 2014). This uncertainty is a source of insecurity among cocoa farmers, for their cocoa trees and for other trees as well. The power dynamics within the cocoa producing communities, where village chiefs have the right and the duty to (re)allocate land for different kinds of development and may even, at least de facto, give external parties short-term user rights to farmers' land, affect not only farmers' choices vis-à-vis agroforestry practices, but also buying companies' sustainability projects. Many if not all international cocoa traders are implementing projects in the cocoa producing areas in Ghana with the stated aim to increase cocoa production, improve farmer livelihoods and build climate resilience among producers—by handing out tree seedlings and training farmers in shade tree management (Carodenuto & Buluran, 2021; Thorlakson, 2018). As project participation may give preferential access to training, inputs and other kinds of support, the power dynamics within cocoa communities are instrumental in determining who will be able to engage in sustainability projects. This may lead to marginalized producers, e.g., descendants of migrants, being excluded from potential project benefits and pushed to even more disadvantaged situations. A similar scenario may play out for female farmers, who despite performing half of the work on cocoa farms are vastly underrepresented among the officially registered cocoa farmers due to registration being tied to land tenure systems that traditionally favor men (Barrientos & Bobie, 2016).

The need to remove risks and uncertainties from the shoulders of farmers is clear, not least for the facilitation of agroforestry promotion. Given the long tradition of traditional land authorities and the numerous actors involved in land governance, it will be no easy feat to enhance the transparency and accountability of these institutions, though this is called for to increase land security (Kasanga & Ashie Kotey, 2001). The overlapping and sometimes competing rights in administration systems for trees and land should be integrated so trees are tied to the farmland, affording all tree tenure rights to the landholder, when relevant, under the same conditions as those pertaining to food and cash crops. This would remove the administrative burden and costs of tree registration in both public and private programs.

Additionally, to increase adoption of shade trees, it is necessary to improve the current tree seedling distribution by COCOBOD's Seed Production Division, which is currently limited by farmers having to cover transportation costs. The program is essentially funded by the cocoa sector, including the farmers, through COCOBOD's price regulation and the proceeds of the cocoa export. It is by no means an easy task as the current mass spraying programs are already flawed, as reported by farmers. With tree seedling distribution becoming widespread also in corporate extension programs, there is an opportunity for public–private partnerships in a commercially pre-competitive setup including decentralized nurseries and strengthened distribution channels. Falling short of unifying tree and farmland tenure, the registration of newly planted trees should be an integrated part of tree seedling distribution programs, e.g., by digital receipts registered with farmers or farmer organizations upon delivery of the trees. This setup could piggyback on the registration of farmers' passbooks that have shown to work well for registering cocoa production in each cocoa district. Furthermore, it is important that a greater variety of tree seedlings is distributed through these programs. These programs should consider both farmers' preferred shade tree species and location-specific factors that influence the cocoa agroforestry system, such as the local climate and climate change predictions.

The management of shade trees may not be a panacea for decent cocoa-based livelihoods and a living income for farmers. However, when implemented on sound management practices and based on secure rights to land, cocoa and shade trees, agroforestry has the potential to generate diverse income streams for farming households, provide ecosystem services at the societal level, improve climate resiliency and supply cocoa raw materials to a global consumer base.

4.5   Conclusion

From pre-Hispanic Mayan cultivation of cocoa to present-day cocoa farms in Ghana, the farming of cocoa is more than the sole marketable value of the cocoa beans. While Ghanaian farmers do not attribute ceremonial values to their cocoa trees like the Mayans do, they do derive non-cocoa values from the cocoa plots, especially when managed as agroforestry systems. Ecosystem goods and services are provided by the shade trees and the shady environment to the farming households, such as food, fodder, medicine and materials. Trees are seen by farmers as increasingly important given their recent experiences of a warming climate, both for adapting to droughts and higher temperature and for mitigating further climate change. However, by focusing only on the apparent synergies between climate change resilience and farmer benefits from agroforestry, it is easy to overlook institutional factors that can prevent cocoa farmers from engaging in longer-term agroforestry practices and thereby benefiting from the opportunities they present. Especially, the institutional complex surrounding land and tree tenure creates high uncertainties for farmers regarding their ability to enjoy the benefits from their shaded cocoa plots. The costly registration of trees with Forestry authorities leads to limited user rights to trees on cocoa farmland, removing the economic incentives to care for trees. For some farmers, the risks of the loss of cocoa plots to mining activities, at the discretion of village chiefs, add additional insecurity to cocoa-based livelihoods and thus to longer-term investments in trees. While major land reforms may not be on the horizon, there is a need to unify tree and land rights systems to avoid overlapping and conflicting tenure regimes. This will ease current struggles among both private and public programs for tree seedling dissemination and the promotion of agroforestry.

REFERENCES

Abdulai, I., Vaast, P., Hoffman, M., Asare, R., Jassogne, L., Asten, V. P., Rotter, P. R., & Graefe, S. (2018). Cocoa agroforestry is less resilient to sub-optimal and extreme climate than cocoa in full sun. *Global Change Biology, 24*(1), 273–286. https://doi.org/10.1111/gcb.14044

Abou Rajab, Y., Leuschner, C., Barus, H., Tjoa, A., & Hertel, D. (2016). Cacao cultivation under diverse shade tree cover allows high carbon storage and sequestration without yield losses. *PLoS ONE, 11*(2), e0149949. https://doi.org/10.1371/journal.pone.0149949

Acheampong, E. O., Macgregor, C. J., Sloan, S., & Sayer, J. (2019). Deforestation is driven by agricultural expansion in Ghana's forest reserves. *Scientific African, 5*, e00146. https://doi.org/10.1016/j.sciaf.2019.e00146

Ahenkan, A., & Boon, E. (2011). Improving nutrition and health through non-timber forest products in Ghana. *Journal of Health, Population, and Nutrition, 29*(2), 141–148. https://doi.org/10.3329/jhpn.v29i2.7856

Amanor, S. K. (1996). *Managing trees in the farming system: The perspectives of farmers* (Forest Farming Series No. 1; 202 pp.). Forestry Department, Ghana.

Anglaaere, L. C. N., Cobbina, J., Sinclair, F. L., & McDonald, M. A. (2011). The effect of land use systems on tree diversity: Farmer preference and species composition of cocoa-based agroecosystems in Ghana. *Agroforestry Systems, 81*, 249–265. https://doi.org/10.1007/s10457-010-9366-z

Asare, R., Markussen, B., Asare, R. A., Anim-Kwapong, G., & Ræbild, A. (2019). On-farm cocoa yields increase with canopy cover of shade trees in two agro-ecological zones in Ghana. *Climate and Development, 11*(5), 435–445. https://doi.org/10.1080/17565529.2018.1442805

Asitoakor, B. K., Vaast, P., Ræbild, A., Ravn, H. P., Eziah, V. Y., Owusu, K., Mensah, E. O., & Asare, R. (2022). Selected shade tree species improved cocoa yields in low-input agroforestry systems in Ghana. *Agricultural Systems, 202*(103476), 1–9. https://doi.org/10.1016/j.agsy.2022.103476

Awuah, R., & Kyereh, B. (2019). How farmers develop local ecological knowledge for on-farm tree management: The perspectives of some farming communities of Ghana. *Natural Resources Forum, 44*, 287–383. https://doi.org/10.1111/1477-8947.12210

Barrientos, S., & Bobie, A. O. (2016). *Promoting gender equality in the cocoa-chocolate value chain: Opportunities and challenges in Ghana* (GDI Working Paper 2016-006). University of Manchester.

Bentley, J. W., Boa, E., & Stonehouse, J. (2004). Neighbor trees: Shade, inter-cropping, and cacao in Ecuador. *Human Ecology, 32*, 241–270. https://doi.org/10.1023/B:HUEC.0000019759.46526.4d

Boadi, S. A., Olwig, M. F., Asare, R., Bosselmann, A. S., & Owusu, K. (2022). The role of innovation in sustainable cocoa cultivation: Moving beyond mitigation and adaptation. In M. Coromaldi & S. Auci (Eds.), *Climate-induced innovation: Mitigation and adaptation to climate change* (pp. 47–80). Springer.

Bos, M. M., Steffan-Dewenter, I., & Tscharntke, T. (2007). Shade tree management affects fruit abortion, insect pests and pathogens of cacao. *Agriculture, Ecosystems and Environment, 120*(2–4), 201–205. https://doi.org/10.1016/j.agee.2006.09.004

Bunn, C., Läderach, P., Quaye, A., Muilerman, S., Noponen, M. R. A., & Lundy, M. (2019, June). Recommendation domains to scale out climate change adaptation in cocoa production in Ghana. *Climate Services, 16*. https://doi.org/10.1016/j.cliser.2019.100123

Carodenuto, S., & Buluran, M. (2021). The effect of supply chain position on zero-deforestation commitments: Evidence from the cocoa industry. *Journal of Environmental Policy & Planning*. https://doi.org/10.1080/1523908X.2021.1910020

Cerda, R., Deheuvels, O., Calvache, D., Niehaus, L., Saenz, Y., Kent, J., Vilchez, S., Villota, A., Martinez, C., & Somarriba, E. (2014). Contribution of cocoa agroforestry systems to family income and domestic consumption: Looking toward intensification. *Agroforestry Systems, 88*(6), 957–981. https://doi.org/10.1007/s10457-014-9691-8

Coq-Huelva, D., Higuchi, A., Alfalla-Luque, R., Burgos-Morán, R., & Arias-Gutiérrez, R. (2017). Co-evolution and bio-social construction: The Kichwa Agroforestry Systems (Chakras) in the Ecuadorian Amazonia. *Sustainability, 9*(10), 1920. https://doi.org/10.3390/su9101920

Curry, G. N., Koczberski, G., Lummani, J., Nailina, R., Peter, E., McNally, G., & Kuaimba, O. (2015). A bridge too far? The influence of socio-cultural values on the adaptation responses of smallholders to a devastating pest outbreak in cocoa. *Global Environmental Change, 35,* 1–11. https://doi.org/10.1016/j. gloenvcha.2015.07.012

FAO. (2009). *Food security and agricultural mitigation in developing countries: Options for capturing synergies.* Food and Agriculture Organization of the United Nations.

Franzen, M., & Borgerhoff Mulder, M. (2007). Ecological, economic and social perspectives on cocoa production worldwide. *Biodiversity and Conservation, 16,* 3835–3849. https://doi.org/10.1007/s10531-007-9183-5

Friedman, R., Hirons, M. A., & Boyd, E. (2018). Vulnerability of Ghanaian women cocoa farmers to climate change: A typology. *Climate and Development, 11.* https://doi.org/10.1080/17565529.2018.1442806

Gaither, C. J., Yembilah, R., & Samar, S. B. (2019). Tree registration to counter elite capture of forestry benefits in Ghana's Ashanti and Brong Ahafo regions. *Land Use Policy, 85,* 340–349. https://doi.org/10.1016/j.landusepol.2019. 04.006

Gockowski, J., Tchatat, M., Dondjang, J.-P., Hietet, G., & Fouda, T. (2010). An empirical analysis of the biodiversity and economic returns to cocoa agroforests in Southern Cameroon. *Journal of Sustainable Forestry, 29*(6–8), 638–670. https://doi.org/10.1080/10549811003739486

Gómez-Pompa, A., Flores, J. S., & Fernández, M. A. (1990). The sacred cacao groves of the Maya. *Latin American Antiquity, 1*(3), 247–257. http://www. jstor.com/stable/972163

Graefe, S., Meyer-Sand, L., Chauvette, K., et al. (2017). Evaluating farmers' knowledge of shade trees in different cocoa agro-ecological zones in Ghana. *Human Ecology, 45,* 321–332. https://doi.org/10.1007/s10745-017-9899-0

Herzog, F. (1994). Multipurpose shade trees in coffee and cocoa plantations in Côte d'Ivoire. *Agroforest Systems, 27,* 259–267. https://doi.org/10.1007/BF00705060

Hirons, M., McDermott, C., Asare, R., Morel, A., Robinson, E., Mason, J., Boyd, E., Malhi, Y., & Norris, K. (2018). Illegality and inequity in Ghana's cocoa-forest landscape: How formalization can undermine farmers control and benefits from trees on their farms. *Land Use Policy, 76,* 405–413. https://doi.org/10.1016/j.landusepol.2018.02.014

Jamal, A. M., Antwi-Agyei, P., Baffour-Ata, F., Nkiaka, E., Antwi, K., & Gbordzor, A. (2021). Gendered perceptions and adaptation practices of smallholder cocoa farmers to climate variability in the Central Region of Ghana. *Environmental Challenges, 5,* 100293. https://doi.org/10.1016/j. envc.2021.100293

Kaba, J. S., Otu-Nyanteh, A., & Abunyewa, A. A. (2020). The role of shade trees in influencing farmers' adoption of cocoa agroforestry systems: Insight from semi-deciduous rain forest agroecological zone of Ghana. *NJAS—Wageningen*

Journal of Life Sciences, 92, 100332. https://doi.org/10.1016/j.njas.2020.
100332

Kalischek, N., Lang, N., Renier, C., Caye Daudt R., et al. (2022). *Satellite-based high-resolution maps of cocoa for Côte d'Ivoire and Ghana.* arXiv:2206.061 19v2

Kasanga, K., & Ashie Kotey, N. (2001). *Land management in Ghana: Building on tradition and modernity.* International Institute for Environment and Development.

Koko, L. K., Snoeck, D., Lekadou, T. T., & Assiri, A. A. (2013). Cacao-fruit tree intercropping effects on cocoa yield, plant vigour and light interception in Côte d'Ivoire. *Agroforestry Systems, 87*, 1043–1052. https://doi.org/10.1007/s10457-013-9619-8

Kufer, J., Grube, N., & Heinrich, M. (2006). Cacao in Eastern Guatemala—A sacred tree with ecological significance. *Environment, Development and Sustainability, 8*, 597–608. https://doi.org/10.1007/s10668-006-9046-3

Mayer, J., & Ashie Kotey, E. N. (1996). *Local institutions and adaptive forest management in Ghana* (IIED Forestry and Land Use Series no. 7). International Institute for Environment and Development.

Nasser, F., Maguire-Rajpaul, V. A., Dumenu, W. K., & Wong, G. Y. (2020). Climate-smart cocoa in Ghana: How ecological modernisation discourse risks side-lining cocoa smallholders. *Frontiers in Sustainable Food Systems, 4*, 73. https://doi.org/10.3389/fsufs.2020.00073

Nunoo, I., & Owusu, V. (2017). Comparative analysis on financial viability of cocoa agroforestry systems in Ghana. *Environment, Development and Sustainability, 19*(1), 83–98. https://doi.org/10.1007/s10668-015-9733-z

Obeng, E. A., Obiri, B. D., Oduro, K. A., Pentsil, S., Anglaaere, L. C., Foli, E. G., & Ofori, D. A. (2020). Economic value of non-market ecosystem services derived from trees on cocoa farms. *Current Research in Environmental Sustainability, 2*, 100019.

Obiri, B. D., Bright, G. A., McDonald, M. A., Anglaaere, L. C. N., & Cobbina, J. (2007). Financial analysis of shaded cocoa in Ghana. *Agroforestry Systems, 71*, 139–149. https://doi.org/10.1007/s10457-007-9058-5

Ofori-Bah, A., & Asafu-Adjaye, J. (2011). Scope economies and technical efficiency of cocoa agroforesty systems in Ghana. *Ecological Economics, 70*, 1508–1518. https://doi.org/10.1016/j.ecolecon.2011.03.013

Ofosu, G., & Sarpong, D. (2022). Mineral exhaustion, livelihoods and persistence of vulnerabilities in ASM settings. *Journal of Rural Studies, 92*, 154–163. https://doi.org/10.1016/j.jrurstud.2022.03.029

Okuh, G., & Hilson, G. (2011). Poverty and livelihood diversification: Exploring the linkages between smallholder farming and artisanal mining in rural Ghana. *Journal of International Development, 23*, 1100–1114. https://doi.org/10.1002/jid.1834

Oladokun, M. A. O. (1990). Tree crop based agroforestry in Nigeria: A checklist of crops intercropped with cocoa. *Agroforest Systems, 11*, 227–241. https://doi.org/10.1007/BF00045901

Ordway, E. M., Asner, G. P., & Lambin, E. F. (2017). Deforestation risk due to commodity crop expansion in sub-Saharan Africa. *Environmental Research Letters, 12*, 044015.

Orozco-Aguilar, L., López-Sampson, A., Leandro-Muñoz, M. E., Robiglio, V., Reyes, M., Bordeaux, M., Sepúlveda, N., & Somarriba, E. (2021). Elucidating pathways and discourses linking cocoa cultivation to deforestation, reforestation, and tree cover change in Nicaragua and Peru. *Frontiers in Sustainable Food Systems, 5*, 635779. https://doi.org/10.3389/fsufs.2021.635779

O'Sullivan, R., Roth, M., Antwi, Y. A., Ramirez, P., & Sommerville, M. (2018, March). *Land and tree tenure innovations for financing smallholder cocoa rehabilitation in Ghana*. Winrock International. Paper presented at 2018 World Bank conference on land and poverty, Washington.

Quisumbing, A., Aidoo, J. B., Payongayong, E., & Otsuka, K. (2001). Agroforestry management in Ghana. In K. Otsuka & F. Place (Eds.), *Land tenure and natural resource management* (pp. 53–96). John Hopkins University Press.

Rice, R. A., & Greenberg, R. (2000). Cacao cultivation and the conservation of biological diversity. *AMBIO: A Journal of the Human Environment, 29*(3), 167–173.

Ruf, F., & Schroth, G. (2004). Chocolate forests and monocultures: A historical review of cocoa growing and its conflicting role in tropical deforestation and forest conservation. In G. Schroth, G. A. B. Da Fonseca, H. Celia, C. Gascon, H. Vasconcelos, & A.-M. Izac (Eds.), *Agroforestry and biodiversity conservation in tropical landscapes* (pp. 107–134). Island Press.

Ryan, Ó. (2011). *Chocolate nations: Living and dying for cocoa in West Africa*. Zed Books.

Schulz, B., Becker, B., & Götsch, E. (1994). Indigenous knowledge in a 'modern' sustainable agroforestry system—A case study from eastern Brazil. *Agroforest Systems, 25*, 59–69. https://doi.org/10.1007/BF00705706

Siaw, D., Ofuso, G., & Sarpong, D. (2023). Cocoa production, farmlands, and the galamsey: Examining current and emerging trends in the ASM-agriculture nexus. *Journal of Rural Studies, 101*, 103044. https://doi.org/10.1016/j.jrurstud.2023.103044

Smith Dumont, E., Gnahoua, G. M., Ohouo, L., et al. (2014). Farmers in Côte d'Ivoire value integrating tree diversity in cocoa for the provision of ecosystem services. *Agroforest Systems, 88*, 1047–1066. https://doi.org/10.1007/s10457-014-9679-4

Spichiger, R., & Stacey, P. (2014). *Ghana's land reform and gender equality* (DIIS Working Paper 2014:01). Danish Institute for International Studies.

Steinberg, M. K. (2002). The globalization of a ceremonial tree: The case of cacao (*Theobroma cacao*) among the Mopan Maya. *Economic Botany, 56*, 58–65.

Thorlakson, T. (2018). A move beyond sustainability certification: The evolution of the chocolate industry's sustainable sourcing practices. *Business Strategy and the Environment, 27*, 1653–1665.

Vaast, P., Harmand, J.-M., Rapidel, B., Jagoret, P., & Deheuvels, O. (2015). Coffee and cocoa production in agroforestry—A climate-smart agriculture model. In E. Torquebiau (Ed.), *Climate change and agriculture worldwide*. Springer. https://doi.org/10.1007/978-94-017-7462-8_16

Zarrillo, S., Gaikwad, N., Lanaud, C., et al. (2018). The use and domestication of Theobroma cacao during the mid-Holocene in the upper Amazon. *Nature Ecology and Evolution, 2*, 1879–1888. https://doi.org/10.1038/s41 559-018-0697-x

Open Access This chapter is licensed under the terms of the Creative Commons Attribution 4.0 International License (http://creativecommons.org/licenses/by/4.0/), which permits use, sharing, adaptation, distribution and reproduction in any medium or format, as long as you give appropriate credit to the original author(s) and the source, provide a link to the Creative Commons license and indicate if changes were made.

The images or other third party material in this chapter are included in the chapter's Creative Commons license, unless indicated otherwise in a credit line to the material. If material is not included in the chapter's Creative Commons license and your intended use is not permitted by statutory regulation or exceeds the permitted use, you will need to obtain permission directly from the copyright holder.

CHAPTER 5

Household Economics of Cocoa Agroforestry: Costs and Benefits

Sylvester Afram Boadi, *Aske Skovmand Bosselmann*,
Kwadwo Owusu, *Richard Asare*, *and Mette Fog Olwig*

Abstract Current research suggests that cocoa agroforestry systems could offer stable yields, additional benefits and income from shade trees, despite potential added costs, such as from the purchase of insecticides. There is a paucity of profitability studies of different cocoa agroforestry systems. Only few of them go beyond a narrow focus on cocoa yields to model the entire agroforestry system and thus do not advance our understanding of the socio-economic value of other ecosystem goods. Based on

S. A. Boadi (✉) · K. Owusu
Department of Geography and Resource Development, University of Ghana, Accra, Ghana
e-mail: saboadi@csir.org.gh

K. Owusu
e-mail: kowusu@ug.edu.gh

S. A. Boadi · A. S. Bosselmann
Department of Food and Resource Economics, University of Copenhagen, Frederiksberg, Denmark
e-mail: ab@ifro.ku.dk

© The Author(s) 2024
M. F. Olwig et al. (eds.), *Agroforestry as Climate Change Adaptation*,
https://doi.org/10.1007/978-3-031-45635-0_5

survey data covering a thousand cocoa plots and group interviews with cocoa farmers, we explore the costs and benefits at the household level of including trees in cocoa systems. Comparing low and medium tree diversity systems, we find that income from cocoa beans, timber and fruit trees are higher and labour costs are lower in plots with medium diversity, while insecticide costs are lower on low-diversity plots. Overall, net benefits were higher on cocoa plots with higher tree diversity. Thus, cocoa agroforestry systems offer cost-reduction and income-improving advantages. Since cocoa systems vary among different agro-ecological zones in Ghana, we recommend that interventions aimed at increasing tree diversity consider the specific management practices of each farming household and the location in question.

Keywords Cocoa agroforestry · Tree diversity · Household economics · Profitability · Cost reduction · Income diversification

5.1   INTRODUCTION

Approximately two million households depend on cocoa farming as their primary source of livelihood income in West Africa (World Cocoa Foundation, 2022). For these farmers, cocoa field productivity is of utmost importance. Traditional pathways to increasing cocoa yields across the West African cocoa belt have involved the expansion of farms into forest areas and the cultivation of hybrid seeds mostly under full-sun systems, i.e. in monocrop systems with little to no shade. Since 1950, Côte d'Ivoire

---

S. A. Boadi
CSIR-Water Research Institute, Accra, Ghana

R. Asare
International Institute of Tropical Agriculture (IITA), Accra, Ghana
e-mail: r.asare@cgiar.org

M. F. Olwig
Department of Social Sciences and Business, Roskilde University, Roskilde, Denmark
e-mail: mettefo@ruc.dk

and Ghana have lost, respectively, 90 and 65% of their forest areas, mainly due to agricultural expansion with cocoa as the dominant crop (Kalischek et al., 2022). Deforestation driven by cocoa continues (Barima et al., 2016; Nunoo et al., 2015; Ongolo et al., 2018), including in forest reserves and national parks, where small-scale farming of cocoa is among the leading causes of deforestation and forest degradation (Acheampong et al., 2019; Kalischek et al., 2022). This strategy for cocoa expansion is unsustainable and there is an urgent need to identify ways to increase production without compromising environmental sustainability. Environmental sustainability is high on the agenda among major cocoa-buying companies (Carodenuto & Buluran, 2021), if not for the sake of the forest, then for the sake of market access, as the world's major chocolate consuming regions, the EU and the United States, are developing new regulations that are expected to hinder cocoa imports unless documented as deforestation-free. Cocoa agroforestry has been presented by researchers and farmers as a more sustainable and climate-resilient pathway for maintaining and even increasing cocoa farm outputs (Daghela Bisseleua, 2019).

As discussed in more detail in the previous chapters, cocoa agroforestry involves planting, or managing the regeneration of companion trees and/ or crops with cocoa for agronomic, environmental and economic benefits (Asare, 2006; Asare & Asare, 2008). Cocoa agroforestry has been shown to enhance soil health, improve climate resilience, sequester carbon, increase farmer income, secure household food and nutrition needs, reduce pest and disease outbreaks (by acting as barriers) and improve biodiversity (Blaser et al., 2017; Nair & Nair, 2014). Yet, as this chapter will show, the types and extent of benefits to specific farms vary depending on the particular farming systems employed, including how much labour and inputs are used, but also tree species diversity. Thus, while the trade-off and cost aspects of managing cocoa agroforestry systems are different from full-sun systems, they also vary between different cocoa agroforestry arrangements (Asare et al., 2014; Nunoo & Owusu, 2017; Obiri et al., 2007).

Historically, West African cocoa farmers increased cocoa bean production by shifting to new cultivation frontiers, which had several advantages compared to other methods. Cocoa-cultivation frontiers are geographical regions of abundant unoccupied land resources (often forest land) that did not previously include cocoa cultivation. As a result of the introduction of cocoa cultivation, these regions have experienced substantial flows of people (migration) and capital (Knudsen & Agergaard, 2015). The expansion into new frontiers was mostly a cost-saving strategy, e.g. to resolve ecological instability in current production areas, such as helping manage pest and disease infestations and declining soil fertility (Kolavalli & Vigneri, 2011). This practice predominated because of the availability of large expanses of forest land in cocoa-cultivation belts (Asare, 2005; Kolavalli & Vigneri, 2011). Across the two leading cocoa-producing countries, Côte d'Ivoire and Ghana, land has become scarce, and farmlands fragmented. In Ghana, there are no major cultivation frontiers left (Amanor et al., 2021; Asare & Ræbild, 2016). As a result, farmers turned to intensive full-sun cultivation, which allows for increasing productivity through appropriate agricultural practices and the rational application of agrochemicals (fungicides and insecticides) and fertilizers. However, as the climate changes and becomes warmer or wetter, the pressure of diseases and pests on cocoa trees in full-sun cultivation may increase, leading to higher costs of production and lower income (Asare et al., 2014; Schroth et al., 2000). In contrast, several of the potential benefits of cocoa agroforestry practices, such as improving soil fertility, reducing pest infestations, reducing weed growth and moderating the impacts of dry spells and drought on yields (Abou Rajab et al., 2016; Bos et al., 2007; Tscharntke et al., 2011), all mitigate the impacts of climate change while reducing costs for farmers (Cerda et al., 2014).

In Ghana, cocoa farming serves as the main livelihood option for about 550,000 smallholder households (Ghana Statistical Service, 2019), whose livelihoods depend directly on cocoa farm yields. In such households, earnings from cocoa bean sales are a key component of total household incomes and critical to meeting household needs related to food, health, education and other necessities. However, cocoa farming is an input-intensive activity whose benefits mostly depend on how much labour, agrochemicals and fertilizer producers apply (Asare et al., 2019). Prior to the individualization of labour due to increasing urbanization and commercialization across the West African cocoa-growing belt, family, neighbours and community members were a key source of unpaid labour for cocoa cultivation that kept costs down. However, with increasing out-migration of labour from the cocoa areas and the proliferation of other competing economic activities in the cocoa-growing areas, such as small-scale and large-scale mining and sand winning, labour for cocoa farming has become scarce, increasing its cost (Ministry of Manpower Youth & Employment, 2007). This shortage of labour has furthermore been cited as a factor in the problem of child labour in cocoa cultivation in Ghana and across West Africa (Sadhu et al., 2020). The rising costs of labour also reduce the amount of money farmers can spend on fertilizers and agrochemicals such as herbicides, fungicides and insecticides. As a result, the inputs farmers use decrease, and by extension the outputs and cocoa farm incomes. The cost-reduction advantages hypothesized for cocoa agroforestry systems could therefore be helpful to farmers, especially if these were also associated with increases in farm outputs and benefits.

Based on a cost–benefit analysis, this chapter explores whether smallholder cocoa farmers increase their incomes and improve their livelihoods when implementing tree diverse cocoa-cultivation systems focusing on the effect of the level of tree species diversity. Specifically, the chapter asks the following questions: (i) how does the level of tree species diversity affect the costs and benefits of cocoa agroforestry? And (ii) what are the household economic implications of managing cocoa agroforestry systems across different climate gradients in Ghana? These questions are important in an era when sustainable production has become more important for conserving resources, improving the climate resilience of agricultural systems and enhancing livelihoods.

The chapter is organized into four sections. Following the introduction, Sect. 5.2 provides a review of the larger literature on cocoa farmers' livelihoods and the role of agroforestry in safeguarding livelihoods, while Sects. 5.3 and 5.4 detail the conceptual framework and methodology. Section 5.5 presents the results and discusses policy implications based on the findings, while Sect. 5.6 provides a conclusion.

5.2 Literature Review: Cocoa Farmers' Livelihoods and the Role of Agroforestry

The integration of trees into cropping systems provides smallholders with alternative livelihoods and income besides earnings from the sale of cocoa beans (Atangana et al., 2014; Cerda et al., 2014; Graefe et al., 2017; Ruf & Schroth, 2004). For example, it provides farmers additional income through the sale of firewood, fruits and, in some instances, timber (Asare et al., 2014; Graefe et al., 2017). The integration of trees into cocoa farms and landscapes is also important because trees help moderate the impact of climatic stress (from higher temperatures and droughts), provide shade for the cocoa trees, serve as barriers to the spreading of pests and diseases, and sequester/store carbon (Abou Rajab et al., 2016; Asare et al., 2014;

Bos et al., 2007; Daghela Bisseleua et al., 2013; Graefe et al., 2017; Smith Dumont et al., 2014; Tscharntke et al., 2011). On the other hand, cocoa agroforestry systems also introduce new costs. For example, shade trees may cause competition for root space and nutrients in young plantations (Smith Dumont et al., 2014; see also Chapter 2 in this volume), while excessive shade may increase pest and disease pressures (Graefe et al., 2017; see also Chapter 3 this volume). Cocoa agroforestry is also associated with a reduction in yields compared to intensified full-sun systems, assuming farmers apply the required inputs, and not considering the impacts of climate change on full-sun systems (Nunoo & Owusu, 2017). However, the actual and potential costs and benefits that smallholders derive from these cocoa agroforestry systems are influenced by institutional, technical, marketing and legal arrangements (Mugure et al., 2013; Roth et al., 2018). In Ghana and Côte d'Ivoire, for example, factors such as land and tree tenure arrangements; whether farmers originate from forest zones or savannah zones; and social networks; largely determine whether farmers integrate trees into their cocoa systems and affect the benefits farmers derive from such cocoa systems (Gyau et al., 2015; Roth et al., 2018; Ruf & Schroth, 2004; see also Chapter 4).

Household socio-economic characteristics, including resource endowments and household assets, are also important in determining the costs and benefits farmers derive from their cocoa systems. Based on socio-economic characteristics, cocoa farmers in Ghana could be grouped into aged and young, rich and poor, educated and illiterate, well-diversified and less-diversified farms, male and female, and indigenes and migrant farmers, among others. Each of these farmer types has different capacities to adopt cocoa agroforestry practices, based on their household socio-economic characteristics and tenure arrangements. For instance, the average age of cocoa farmers in Ghana currently is above fifty (Asamoah et al., 2015), which impacts willingness and ability to adopt innovative and improved cultivation practices, such as cocoa agroforestry (Barrientos et al., 2008; Boadi et al., 2022; Djokoto et al., 2016). Some of these socio-economic dynamics influencing the costs and benefits associated with cocoa agroforestry systems are modulated by prevailing state policies on access to inputs and producer prices.

Several state policies and programmes have been introduced in Ghana over the years to improve cocoa farmers' cultivation practices, leading to potential cost-saving advantages for cocoa farmers as well. The most relevant of these policies for cocoa agroforestry practices are the

Cocoa Mass Spraying Programme introduced in 2001, the Cocoa Hi-Tech Programme introduced in 2002/2003 and the Hand Pollination Programme introduced in 2017 (COCOBOD, 2018; Kolavalli & Vigneri, 2017). The Mass Spraying Programme, which increased the application of insecticides for effective pest and disease control, helped farmers control black pod infestations, which have been linked to the introduction of shade trees on cocoa farms—especially when the shade canopies are not managed well (Bos et al., 2007; Schroth et al., 2000; Tscharntke et al., 2011). The Cocoa Hi-Tech Programme, which involved the distribution of subsidized and/or free fertilizers to farmers (Kolavalli & Vigneri, 2017), increased yields, alleviating farmers' fears of reduced yields from adopting cocoa agroforestry systems. The Hand Pollination Programme has a similar potential to increase yields.

Increases in the cocoa producer price are also important to increase farmers' economic room for improvement in farm management, such as affording to purchase inputs, including planting materials for shade trees. A noteworthy development with significant implications for producer prices in Ghana is the "Living Income Differential Policy" (LID), which aims to reduce the differential between current incomes and the income needed for farmers to live a decent life. In 2019, the governments of Côte d'Ivoire and Ghana formed an alliance to demand that large trading companies and other sector players in Europe and North America pay a LID premium of USD 400 per tonne of cocoa purchased. While not yet implemented fully in price-setting policies, the LID policy has led to increments in purchasing prices in Ghana for a 64-kilogramme (kg) bag of cocoa from Ghana Cedis (GHS) 515 in the 2019/2020 cocoa season, i.e. 87.5 USD per bag (USD 1400 per tonne), to GHS 660 in the 2020/2021 cocoa season, i.e. 102 USD per bag (USD 1632 per tonne). The policy has also led to the International Cocoa Agreement, signed by producing and consuming member countries, which includes "a reference to remunerative prices to reach economic sustainability" and achieve a living income (ICCO, 2022). The international trading companies have supported the LID policy in their communications, but some push-back on the price increases has been seen in the sector, such as buyers' lowering the origin or quality differentials to off-set some of the LID-related price increment. Historically, favourable producer prices have been associated with improved investments in cocoa farms in Ghana (Kolavalli & Vigneri, 2011), and it is therefore likely that increased purchasing prices could have positive implications for farmers' adoption of improved cultivation practices such as cocoa agroforestry practices.

5.3   CONCEPTUAL FRAMEWORK: AGROFORESTRY, FARMER INCOME AND COST–BENEFIT ANALYSIS

This chapter adopts a definition of cocoa agroforestry systems that includes the stage of the cocoa system and the associated species diversity and distribution. This takes into account the dynamic and constantly changing nature of cocoa systems over the life cycle of the cocoa crop in terms of the crops/trees included and their arrangement in the system. Cocoa agroforestry systems are thus defined as a form of tree diversification, which draws agronomic, environmental and economic benefits from strategically integrating suitable and valuable non-cocoa tree species and other plants in time and space (Asare, 2006). The cocoa plots in the survey were therefore classified into different cocoa systems based on the level of tree diversity on the plots. This classification of plots allows for the inclusion and focus on the total benefits of the cocoa agroforestry system, rather than a narrow focus on shade, which is just one of the characteristics or benefits of integrating trees into cocoa plots.

Using the Shannon–Wiener Diversity Index (H′), the plots were classified as either low-diversity plots or medium-diversity plots. No plot had a sufficiently high diversity to meet the threshold to be classified as a high-diversity plot. The low-diversity plots included cocoa plots close to mono-cropping systems, while medium-diversity plots all fall under agroforestry systems. The Shannon–Wiener Diversity index (H′) is recommended for studies where rare and abundant species are expected to be equally important (Morris et al., 2014). This means that plots classified as having greater diversity in this study had a greater number of rare and abundant species and vice versa.

The chapter's empirical analysis uses the cost–benefit analysis model, which has its roots in utilitarianism (Van Wee & Roeser, 2013). Utilitarianism as a decision-making theory is about maximizing the expected utility of a good, project or policy (Eggleston, 2012). This means that cocoa farmers' decisions on what cultivation systems to adopt are influenced by the expected benefits associated with different practices and systems. Within the context of the current study, cocoa farmers make their final decisions on whether to adopt cocoa agroforestry or full-sun cocoa systems by considering the resultant benefit cost ratios, the difference between benefits and costs, and the return on investments. In the current study, the cost dimensions associated with managing cocoa farms in Ghana were split into costs incurred at the household level by the farmer (private costs) and costs incurred by the state in supporting the cocoa sector (social costs). Private costs include farmers' costs for labour, fertilizer, insecticides, etc., for managing the cocoa farm. Social costs include the costs incurred by the state through its free inputs supply and input subsidy programmes. Private benefits include revenue from cocoa bean sales, food crops and ecosystem goods harvested from the cocoa farm. Social benefits cover revenue the state earns from selling Ghana's cocoa on the international market, etc. See Boadi (2021) for more details on the cost–benefit analysis.

5.4   Methods

Based on household surveys ($n = 402$) and focus-group discussions ($n = 20$) in three different climate impact zones, this chapter assesses the costs and benefits of cocoa agroforestry systems and the contributions of these systems to smallholders' livelihoods. Data were collected from cocoa farmers in twelve cocoa communities in seven administrative districts across Ghana's Ashanti, Ahafo, Western North and Western regions, corresponding to three climate zones along a gradient of increasing dryness and higher maximum temperatures from south to north, as well as greater vulnerability to expected climate change (Bunn et al., 2019). These are known as the Cope Zone (most favourable current climate in relation to cocoa and lowest climate vulnerability), Adjust Zone (moderately favourable current climate and moderate climate vulnerability) and Transform Zone (least favourable current climate and highest climate vulnerability) (see Fig. 5.1). Data was collected for two cocoa seasons: the 2015/2016 season, which was affected by a national drought, and the 2017/2018 season, which was characterized as a "normal" season in terms of seasonal weather patterns.

A total of 1040 cocoa plots belonging to 402 smallholder households in Ghana were surveyed, of which 884 were classified into different cocoa agroforestry systems based on the level of diversity on the plots using the Shannon–Wiener Diversity Index (H'). The remaining 156 plots did not have the required details on the species of integrated trees to be classifiable. The data collected through the household surveys include the

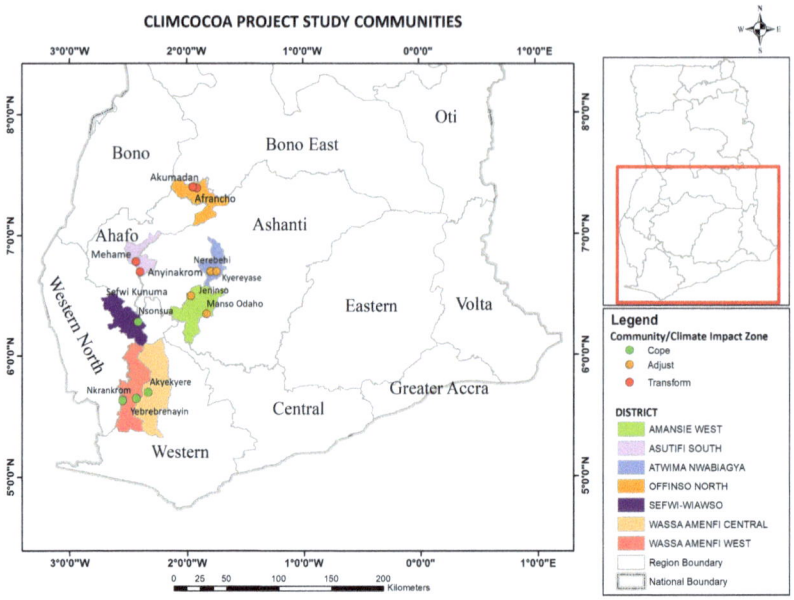

Fig. 5.1 A map of the twelve study communities in cocoa districts in the three climate impact zones (*Source* CLIMCOCOA project)

types, quantities and costs of inputs applied by farmers (land, labour, agrochemicals, capital), cocoa yields, timber and ecosystem goods, such as fruits, firewood and honey harvested from cocoa farms. Using the Shannon–Wiener Diversity Index, 681 cocoa plots were classified as low tree diversity plots and 203 as medium tree diversity plots. The plots were compared using the cost and benefit variables across the different cocoa systems. Cost and benefit parameters for the classified plots were computed using the cost and benefit analysis model described earlier (see Sect. 5.3). The costs of the cocoa systems were derived by calculating all production costs, averaged per hectare, in a complete farming season. This was done for the two seasons of interest. These costs included the cost of labour for weeding, pruning, applying inputs such as fertilizers, insecticides, fungicides, herbicides, harvesting, gathering and breaking pods, fermenting, transporting and drying cocoa beans, cost of inputs and fuel for their application, annual rents on lands where applicable, etc. The cost parameters have been computed in GHS/ha based on the farmers' and state's respective management costs. The benefits of cocoa systems were derived by calculating all the benefits, averaged per hectare, in a complete season. This was also done for the two seasons of interest. The benefits included revenue from cocoa bean sales, including any premiums from certification, food crops, fruits, timber, firewood, honey, mushrooms, bushmeat, snails, fodder/medicinal plants, etc. Similarly, the benefit parameters have been computed in GHS/ha for the farmer and for benefits accruing to the state. The private price for calculating revenue from cocoa beans is derived from the producer prices paid to farmers in the two seasons of interest, while the social price is derived from the FOB price Ghana receives from forward-selling cocoa in the two seasons under consideration. This was derived by converting the FOB price per tonne of cocoa beans to its kg equivalent. These costs and benefits were computed for the low and medium tree diversity cocoa plots for the two seasons and used to derive profitability on a per hectare basis.

5.5   Results and Discussion

The first results section presents findings pertaining to yields (which are linked to benefits) and dominant management practices (which are linked to costs) in low-diversity vs. medium-diversity cocoa plots, as well as in the three different climate zones. These management practices include use of fertilizer, fungicide, insecticide, herbicide and labour. The second section presents cost–benefit analyses of low- versus medium-diversity cocoa plots.

5.5.1   Yields and Household Management Practices

The study found that there were no significant differences in dry cocoa bean yields per hectare between plots of different tree diversity levels. The cocoa bean yield on low-diversity plots was 351.6 kg/ha compared to 358.5 kg/ha on medium-diversity plots (Table 5.1). This finding differs from earlier studies that have found somewhat negative correlations between high levels of shade and yield. This may be due to none of the plots in this study having a high level of tree diversity, thus indicating medium levels of shade. Another reason for this difference is that previous studies, such as that by Nunoo and Owusu (2017), classified cocoa plots using differences in shade levels rather than tree diversity. Nunoo and Owusu (2017) found significant differences in yields of dry cocoa beans per hectare between different levels of shade and reported yields of 516, 588 and 559 kg/ha for plots with no shade, low shade and medium shade (up to 15 shade trees and 85% shade canopy cover) and just 380 kg/ha for high shade cocoa systems (more than 15 trees/ha, with greater than 85% shade canopy cover). These differences illustrate the importance of looking not just at shade levels, but also at cocoa plot tree composition and diversity. Aside from the differences in systems compared and levels of analysis, other studies have attributed yield variations at the plot and farm levels to differences in productive efficiency and in farmers' adoption and use of innovation and technology (Armengot et al., 2020; Meijer et al., 2015).

The results from the analyses of the survey data show that medium tree diversity cocoa plots were significantly larger than low tree diversity plots, indicating that farmers with more resources are more likely to implement agroforestry systems (Table 5.1). Again, this diverges from the findings of Nunoo and Owusu (2017), who found no significant differences in

Table 5.1 Summary statistics associated with surveyed households and cocoa plots

Characteristics	Low diversity	Medium diversity	Mean difference (Low − Medium)
Cocoa plot size (hectares)	1.833	2.412	−0.579*
Yield (kg/ha)	351.576	358.475	−6.899
Total household cocoa landholding (hectares)	6.419	7.236	−0.817
Age of household head (years)	53.562	52.951	0.611
Farming experience of household head (years)	24.179	24.685	−0.506
Household size	5.249	5.532	−0.283
Granular fertilizer (kg/ha)	49.279	46.860	2.419
Foliar fertilizer (litres/ha)	0.731	0.993	−0.262
Fungicide 1 (grams/ha)	759.927	969.867	−209.94
Fungicide 2 (litres/ha)	0.148	0.066	0.082
Insecticide (litres/ha)	2.835	4.510	−1.675*
Herbicide (litres/ha)	1.251	1.418	−0.167
Labour (hours/ha)	770.038	745.954	24.084

*Significant at 5%
Source Survey data from fieldwork in all three climate zones combined

farm sizes when focusing on shade levels rather than tree species diversity. There were no significant differences in total cocoa landholdings for households, the age of household heads, their farming experience, or household size for plots of different tree diversity levels (Table 5.1). Use of insecticides was significantly higher on plots with medium diversity, indicating either that more insecticides were needed or that the farmers with more tree diversity on their plots, which were also significantly larger, had more resources to purchase and apply insecticides. Sellare et al. (2020) also found significantly higher insecticide application on Fairtrade certified cocoa plots compared to uncertified plots in Côte d'Ivoire, attributing the higher use to farmers' extra income from premiums and related services offered by certified cooperatives, allowing them to buy and/or access additional inputs.

The application of fertilizer and agrochemical inputs between low tree diversity and medium tree diversity plots was found not to be significantly different (Table 5.1). A higher fertilizer use was expected on low-diversity plots, which had fewer trees improving the soil fertility. However, across all plots fertilization was well below national recommended levels, which explains the lack of difference. Labour hours per hectare per year, which included both hired and unpaid labour sources (family labour, communal pooled labour and all other unpaid labour used on cocoa plots), was slightly higher on low tree diverse plots than more tree diverse plots, although the differences were not statistically significant.

As shown in Table 5.2, cocoa bean yield per hectare was significantly higher in the Cope zone (423.0 kg/ha) compared to the Adjust (343.04 kg/ha) and Transform (317.87 kg/ha) zones, illustrating the different suitability of the three climate zones for cocoa farming. The national average yield of dry cocoa beans per hectare in the 2017/ 2018 cocoa season was about 500 kg/ha according to FAOSTAT data. Total cocoa landholding and household size differed significantly across the three climate impact zones (Table 5.2). The smaller landholdings in the Cope zone compared to the Transform and Adjust zones are due to the scarcity and traditionally high demand for land in Ghana's main cocoa-cultivation zone and the traditional inheritance practice of dividing cocoa plantations among the owner's children (Löwe, 2017). The smaller household sizes in the Adjust Zone may be due to the closer proximity to urban areas, as rural–urban influences on household size tend to result in larger households in rural areas, as shown by national statistics (Ghana Statistical Service, 2019).

Table 5.2 Summary statistics for households associated with surveyed households and cocoa plots in the different climate impact zones

Characteristics	Transform	Adjust	Cope	Mean differences		
				Adjust and transform	Cope and transform	Cope and adjust
Cocoa plot size (hectares)	2.66	2.08	1.55	−0.58*	−1.12*	−0.54*
Yield (kg/ha)	317.87	343.04	422.97	25.17	105.12*	79.95*
Total HH cocoa landholding (hectares)	9.17	6.68	4.87	−2.49*	−4.29*	−1.81*
Age of household head (years)	56.50	56.10	49.50	−0.40	−7.00*	−6.60*
Farming experience of household head (years)	25.87	25.09	22.41	−0.78	−3.46*	−2.68*
Household size	5.34	4.81	5.93	−0.53*	0.60*	1.13*
Granular fertilizer (kg/ha)	100.79	35.93	20.52	−64.87*	−80.27*	−15.40
Foliar fertilizer (litres/ha)	0.90	0.55	0.77	−0.35*	−0.13	0.22
Fungicide 1 (grams/ha)	937.73	641.81	825.35	−295.92*	−112.38	183.54
Fungicide 2 (litres/ha)	0.09	0.07	0.17	−0.02	0.08	0.10
Insecticide (litres/ha)	2.31	4.72	2.19	2.40*	−0.12	−2.53*
Herbicide (litres/ha)	0.41	1.28	2.26	0.87*	1.85*	0.98*
Labour (hours/ha)	546.90	605.85	953.50	58.95	406.6*	347.65*

*Significant at 5% using a t-test
Source Survey data from fieldwork combining both levels of tree diversity

The application of insecticides in their recommended quantities is key to preventing and/or reducing yield losses from insects like mirids, and the recommendation is for farmers to apply insecticides four times a year, in August, September, October and December (Asare, 2014). The application of insecticide is significantly higher in the Adjust zone compared to the Cope and Transform zones (Table 5.2). Cocoa farmers with sufficient financial capacity therefore tend to apply higher amounts of insecticides, even if not up to the required and recommended amounts. This, coupled with the current finding that higher tree diversity plots were associated with significantly higher plot sizes (land resources), suggests that the observed differences in insecticide application are explained by differences in household financial resources. With land becoming increasingly scarce and fragmented across Ghana's cocoa belts (Bymolt et al., 2018), the differences in plot sizes show that medium-diversity plots are managed by better resourced farmers than low-diversity plots. Well-resourced farmers may be able to purchase and apply significantly more insecticide on their plots than farmers whose cocoa plots have low tree diversity.

5.5.2 Farm Costs and Benefits, and Their Economic Implications

Cost and benefit components are key to estimating and comparing the profitability of cocoa systems. However, other factors such as farmer skills and training, access to inputs and good soils, age of cocoa farmers and plots, among a host of other factors, also influence the cost and benefits associated with specific cocoa farms. For cocoa agroforestry systems, tree diversification generally provides additional benefits through, for example, timber, fruits and other products for subsistence use.

Table 5.3 presents the summary statistics for the cost and benefit categories associated with the low- and medium-diversity cocoa systems. The private costs and benefits are those faced by the farmer, while the social costs and benefits (shown in parenthesis in the table) are those experienced by society or the state. For example, the private cost of fertilizer is based on farmers' purchase prices, while the social cost includes state subsidies, which is why social costs are always higher. Fertilizer, insecticide and herbicide expenditures all increase with increasing tree diversity in both cocoa seasons. The differences are largest in the private costs in both relative and absolute terms.

The higher expenditure on insecticides corresponds to the significantly higher insecticide use in the more diversified plots, as described earlier. Insecticide use is naturally related to the presence of insects, which may be higher in more shaded environments (Graefe et al., 2017; Schroth et al., 2000), though this may not be the case if proper agricultural practices are focused on insect prevention (Armengot et al., 2020). Other studies have mentioned higher applications of agrochemicals, and by extension higher input costs, in full-sun or low-shade cocoa systems than on highly shaded systems (Asare et al., 2014, 2019; Obiri et al., 2007). Such high inputs are mainly due to the higher short-term cocoa yields and the need

Table 5.3 Summary statistics of costs and benefits associated with the different cocoa systems (in GHS/ha). Social costs and benefits shown in parenthesis, where relevant

C/B categories	Diversity level	2015/2016		2017/2018	
Private price 64 kg cocoa bag		425		475	
Social price 64 kg cocoa bag		578		535	
		Mean	N	Mean	N
Cost categories					
Fertilizer	Low diversity	14.19* (30.64)	665	36.67 (74.96)	665
	Medium diversity	46.48* (56.61)	202	60.21 (82.91)	202
Fungicide	Low diversity	19.43 (34.97)	533	22.48 (41.88)	533
	Medium diversity	23.11 (37.81)	176	30.08 (50.93)	176
Insecticide	Low diversity	111.83 (100.49)	657	136.49 (154.43)	657
	Medium diversity	190.24 (167.47)	201	221.33 (252.04)	201
Herbicide	Low diversity	41.45 (36.37)	266	57.31 (57.07)	266
	Medium diversity	39.73 (36.82)	102	52.98 (52.48)	102
Fuel	Low diversity	55.62	486	55.17	555
	Medium diversity	62.09	161	61.49	174
Labour (hired)	Low diversity	1177.81	451	1122.74	530
	Medium diversity	896.91	169	935.89	184
Land	Low diversity	31.38	128	31.25	160
	Medium diversity	29.11	73	29.93	78
Benefit categories					
Cocoa bean income	Low diversity	2278.09 (3098.21)	556	2460.99 (2771.85)	598
	Medium diversity	2503.67 (3404.99)	183	2620.04 (2950.99)	197
Food crop income	Low diversity	138.81	556	160.24	598
	Medium diversity	129.62	183	191.22	197
Certification	Low diversity	–	–	80.47	9
Premium	Medium diversity	–	–	91.85	17
Ecosystem products	Low diversity	–	–	112.73	77
	Medium diversity	–	–	6.89	35
Timber and Fruits	Low diversity	–	–	14.79	662

(continued)

Table 5.3 (continued)

C/B categories	Diversity level	2015/2016		2017/2018	
	Medium diversity	–	–	45.99	197
Profitability category					
Private profitability	Low diversity	1187**		1382**	
	Medium diversity	1526		1658	
Social profitability	Low diversity	2028**		1639**	
	Medium diversity	2454**		1923**	

NB: Ecosystem products include firewood, honey, mushrooms, bushmeat, snails, fodder and medicinal plants
*Significant at 5% between tree diversity levels; **Significant at 5% between years (within tree diversity levels)
Source Survey data from fieldwork

for input intensification in low-shade cocoa systems as a requirement for sustaining yields and managing pests.

Hired labour costs were higher for low tree diversity plots compared to medium tree diversity plots (Table 5.3). The labour was mostly used for manual weeding, which indicates that weeds are a major issue in less diverse and less shaded environments. Regarding the benefit categories, cocoa bean income per hectare increases with increasing tree diversity for both 2015/2016 and 2017/2018 cocoa seasons. Incomes from ecosystem products (i.e. firewood, honey, mushrooms, bushmeat, snails, fodder and medicinal plants) were higher among farmers with low tree diverse cocoa plots, as these products are often collected outside the cocoa farm as well. More than twice the number of farmers with low-diversity plots (77) had income from ecosystem products than farmers with medium-diversity plots (35). Cerda et al. (2014) used cocoa typologies classified according to the size of integrated trees, their densities in the shade canopy and the yields of the agroforestry products, and found cocoa yields made higher contributions to farmer net incomes, with very little contribution from agroforestry products. However, the overall and major contribution of agroforestry products was to household consumption and food security (Cerda et al., 2014). In the current study, medium-diversity plots accrued more incomes from timber and fruits than low-diversity plots. This was expected, as this income category is associated with tree integration, which means that higher tree diversity plots are more likely to contribute higher quantities of these marketable products compared to low tree diverse cocoa systems.

Cocoa bean income per hectare was higher on medium-diversity plots (GHS 2620/ha in the 2017/18 season and GHS 2503/ha in the 2015/ 2016 season) compared to low-diversity plots (GHS 2460/ha in the 2017/2018 season and 2278/ha in the 2015/2016 season). While average private costs did not differ greatly between low- and medium-diversity cocoa plots, the average benefits per hectare were markedly different in both cocoa seasons. In contrast, in their cost–benefit study in Ghana, Nunoo and Owusu (2017) found marked differences in the cost of production between full-sun and cocoa agroforestry systems, with higher total production costs for low-shade cocoa systems and the lowest total production costs for their heavy shade cocoa systems. In Nunoo and Owusu's (2017) study, the difference in cost was because full-sun systems were managed using a high level of inputs.

In sum, this current study found that a greater diversity of tree species could be more profitable than a lower diversity of tree species, despite the higher expenditure on insecticides.

5.5.3   Policy Implications

The findings of this study show that tree integration should be encouraged, especially in cocoa-growing areas, where climate conditions are already dry or projected to become dry, as tree diversity overall increases the profitability and competitiveness of cocoa farms. Yet, it also shows that tree species diversity is important and not just levels of shade. In particular, agroforestry farms with fruit trees are more profitable and more competitive.

One key finding from the chapter is that hired labour costs were higher for low tree diversity cocoa plots compared to medium tree diversity plots. The implication is that cocoa agroforestry systems provide cocoa farmers with an avenue for reducing labour inputs and corresponding costs, e.g. related to manual weeding. Reducing hired labour means that saved costs may be used elsewhere, including to improve cocoa farming. Reduced household labour means more time for other activities, e.g. on-farm or off-farm diversification. The reduced labour demands of agroforestry systems are even more important given recent labour shortages resulting from the proliferation of competing economic activities in cocoa communities, e.g. small-scale illegal mining. As labour costs constitute the greatest percentage of cocoa farmers' total costs, labour-saving practices, associated with higher tree diversity plots in this study, provide avenues for both improving farmers' household incomes and reinvesting the savings from the avoided labour costs back into their farms.

Additionally, the integration of trees and crops that provide marketable products, such as timber and fruit trees, is important for incomes accruing from the agroforestry component of cocoa systems. This requires that the government and other stakeholders in Ghana's cocoa sector take a closer look at the challenges confronting cocoa farmers on restrictions limiting access to relevant tree seedlings and the use of timber trees planted and/ or managed on their farms. On the other hand, low-diversity plots were associated with higher values of so-called environmental incomes, e.g. mushrooms, snails and honey, mainly for subsistence use. This is unexpected and may be due to low-diversity plots belonging more often to smaller farmers with a greater need for the collection of environmental products. Research and policy recommendations on cocoa farmers' choice of cultivation and management practices should thus bear in mind the kinds of benefits different farmers depend on.

Finally, due to variations in levels of resources and social networks, as mentioned above, cocoa farmers' access to and use of inputs such as land, fertilizer and agrochemicals differ. In addition, farmers' use of inputs varies based on differences in access and in how production is managed on plots. These factors will lead to disparate costs and benefits. The findings indicate that farmers who have more resources and better networks, leading to increased access to knowledge of better management practices and ability to afford inputs, choose to implement agroforestry systems. To encourage farmers who do not have these resources or networks to implement agroforestry, it is necessary to provide them with easier access to knowledge of good agricultural management practices, resources and inputs. Importantly, this needs to be tailored to match the specifics of the climate zone and the social position of the farmer.

5.6 Conclusion

The costs and benefits associated with different cocoa agroforestry systems compared to full-sun cocoa systems are important for household economic dynamics in cocoa farming communities. This chapter has explored these dynamics of household economies in Ghana across two broadly defined cocoa farming systems: those with low and high tree diversities respectively. It did so using cost and benefit parameters associated with the cost of inputs at market prices, the corresponding benefits to farmers, the cost of inputs to the economy and the corresponding revenues accruing to the state.

As this study has found, farmers who managed higher tree diversity on plots had correspondingly larger plots than those who managed low-diversity plots. This finding suggests that cocoa farmers' land resources could either directly or indirectly influence the degree and extent of on-farm tree diversification implemented. While certain input costs, such as insecticides, were higher in medium-diversity plots, the labour costs were substantially lower. Gross cocoa bean income per hectare was higher in plots with higher tree diversity, as was the income from timber and fruit trees. Combined, the net benefits favoured cocoa plots with a higher diversity of trees. The major conclusions from the chapter are that cocoa agroforestry systems offer cost-reductions and income-improving advantages and can help cocoa households free up labour for both on-farm and off-farm diversification activities.

Finally, to maximize benefits, recommendations and interventions must be tailored to take into account the specific management practices of each farming household, as well as the climate zone in which the location in question is situated. Additionally, incomes from agroforestry cocoa farms can be improved if restrictions concerning trees planted and managed on farms are addressed.

References

Abou Rajab, Y., Leuschner, C., Barus, H., Tjoa, A., & Hertel, D. (2016). Cacao cultivation under diverse shade tree cover allows high carbon storage and sequestration without yield losses. *PLoS ONE, 11*(2), 1–22. https://doi.org/10.1371/journal.pone.0149949

Acheampong, E. O., Macgregor, C. J., Sloan, S., & Sayer, J. (2019). Deforestation is driven by agricultural expansion in Ghana's forest reserves. *Scientific African, 5*, e00146.

Amanor, K., Yaro, J., & Teye, J. (2021). *Long-term patterns of change in the commercialization of cocoa in Ghana: Forest frontiers and technological transformation* (Working Paper 76; p. 45) Agricultural Policy Research in Africa (APRA).

Armengot, L., Ferrari, L., Milz, J., Velásquez, F., Hohmann, P., & Schneider, M. (2020). Cacao agroforestry systems do not increase pest and disease incidence compared with monocultures under good cultural management practices. *Crop Protection,* *130,* 105047. https://doi.org/10.1016/j.cropro. 2019.105047

Asamoah, M., Aneani, F., Ofori, S., & Branor, P. F. (2015). Analysis of farmers adoption behaviour of CRIG recommended technologies as a package: The case of some self help cocoa farmer associations in the eastern region of Ghana. *Agricultural Sciences,* *6,* 601–608. https://doi.org/10.4236/as. 2015.66059

Asare, R. (2005). *Cocoa agroforests in West Africa: A look at activities on preferred trees in the farming systems.* Forest & Landscape Denmark (FLD).

Asare, R. (2006). *Learning about neighbour trees in cocoa growing systems: A manual for farmer trainers* (Forest & Landscape Development and Environment Working Paper Series no. 4-2006). Forest & Landscape Denmark.

Asare, R. A. (2014). *Understanding and defining climate-smart cocoa: Extension, inputs, yields, and farming practices.* Nature Conservation Research Centre and Forest Trends.

Asare, R., Afari-Sefa, V., Osei-Owusu, Y., & Pabi, O. (2014). Cocoa agroforestry for increasing forest connectivity in a fragmented landscape in Ghana. *Agroforestry Systems,* *88*(6), 1143–1156. https://doi.org/10.1007/s10457-014-9688-3

Asare, R., & Asare, R. A. (2008, September). *A participatory approach for tree diversification in cocoa farms: Ghanaian farmers' experience* (STCP Working Paper Series Issue 9). International Institute of Tropical Agriculture.

Asare, R., Markussen, B., Asare, R. A., Anim-Kwapong, G., & Ræbild, A. (2019). On-farm cocoa yields increase with canopy cover of shade trees in two agro-ecological zones in Ghana. *Climate and Development,* *11*(5), 435–445. https://doi.org/10.1080/17565529.2018.1442805

Asare, R., & Ræbild, A. (2016). Tree diversity and canopy cover in cocoa systems in Ghana. *New Forests,* *47*(2), 287–302. https://doi.org/10.1007/s11056-015-9515-3

Atangana, A., Khasa, D., Chang, S., & Degrande, A. (2014). *Tropical agroforestry.* Springer. https://doi.org/10.1017/CBO9781107415324.004

Barima, Y. S. S., Kouakou, A. T. M., Bamba, I., Sangne, Y. C., Godron, M., Andrieu, J., & Bogaert, J. (2016). Cocoa crops are destroying the forest reserves of the classified forest of Haut-Sassandra (Ivory Coast). *Global Ecology and Conservation,* *8,* 85–98. https://doi.org/10.1016/j.gecco.2016.08.009

Barrientos, S. W., Asenso-Okyere, K., Asuming-Brempong, S., Sarpong, D., Akua Anyidoho, N., Kaplinsky, R., & Leavy, J. (2008). *Mapping sustainable production in Ghanaian Cocoa.* Report to Cadbury Schweppes plc. https://doi.org/10.13140/RG.2.1.4704.4323

Blaser, W. J., Oppong, J., Yeboah, E., & Six, J. (2017). Shade trees have limited benefits for soil fertility in cocoa agroforests. *Agriculture, Ecosystems & Environment,* *243,* 83–91.

Boadi, S. A. (2021). *Socio-economic potential of agroforestry as an alternative livelihood strategy for cocoa farmers in Ghana* (Doctoral dissertation). University of Ghana.

Boadi, S. A., Olwig, M. F., Asare, R., Bosselmann, A. S., & Owusu, K. (2022). The role of innovation in sustainable cocoa cultivation: Moving beyond mitigation and adaptation. In M. Coromaldi & S. Auci (Eds.), *Climate-induced innovation: Mitigation and adaptation to climate change* (pp. 47–80). Springer.

Bos, M. M., Steffan-Dewenter, I., & Tscharntke, T. (2007). Shade tree management affects fruit abortion, insect pests and pathogens of cacao. *Agriculture, Ecosystems and Environment, 120*(2–4), 201–205. https://doi.org/10.1016/j.agee.2006.09.004

Bunn, C., Läderach, P., Quaye, A., Muilerman, S., Noponen, M. R. A., & Lundy, M. (2019, June). Recommendation domains to scale out climate change adaptation in cocoa production in Ghana. *Climate Services, 16*. https://doi.org/10.1016/j.cliser.2019.100123

Bymolt, R., Laven, A., & Tyszler, M. (2018). *Demystifying the cocoa sector in Ghana and Côte d'Ivoire*. Retrieved on 5 May 2020, from https://www.kit.nl/wp-content/uploads/2019/09/Demystifying-cocoa-sector-chapter14-gender-and-cocoa.pdf

Carodenuto, S., & Buluran, M. (2021). The effect of supply chain position on zero-deforestation commitments: Evidence from the cocoa industry. *Journal of Environmental Policy & Planning, 23*(6), 716–731. https://doi.org/10.1080/1523908X.2021.1910020

Cerda, R., Deheuvels, O., Calvache, D., Niehaus, L., Saenz, Y., Kent, J., Vilchez, S., Villota, A., Martinez, C., & Somarriba, E. (2014). Contribution of cocoa agroforestry systems to family income and domestic consumption: Looking toward intensification. *Agroforestry Systems, 88*(6), 957–981. https://doi.org/10.1007/s10457-014-9691-8

COCOBOD. (2018). *Ghana Cocoa Board environmental and social management plan*. COCOBOD.

Daghela Bisseleua, H. (2019). *How Cocoa agroforestry systems can help farmers in West Africa*. World Cocoa Foundation. Retrieved on 6 May 2022, from https://www.worldcocoafoundation.org/blog/how-cocoa-agroforestry-systems-can-help-farmers-in-west-africa/

Daghela Bisseleua, H. B., Fotio, D., Yede, Missoup, A. D., & Vidal, S. (2013). Shade Tree diversity, cocoa pest damage, yield compensating inputs and farmers' net returns in West Africa. *PLoS ONE, 8*(3). https://doi.org/10.1371/journal.pone.0056115

Djokoto, J. G., Owusu, V., & Awunyo-Vitor, D. (2016). Adoption of organic agriculture: Evidence from cocoa farming in Ghana. *Cogent Food & Agriculture, 2*(1), 1242181. https://doi.org/10.1080/23311932.2016.1242181

Eggleston, B. (2012). Utilitarianism. In *Encyclopedia of applied ethics* (2nd ed., Vol. 4, pp. 452–458). Elsevier. https://doi.org/10.1016/B978-0-12-373 932-2.00220-9

Ghana Statistical Service. (2019). *Ghana Living Standards Survey Round 7 (GLSS 7)* (Main report). Ghana Statistical Service.

Graefe, S., Meyer-Sand, L. F., Chauvette, K., Abdulai, I., Jassogne, L., Vaast, P., & Asare, R. (2017). Evaluating farmers' knowledge of shade trees in different cocoa agro-ecological zones in Ghana. *Human Ecology, 45*(3), 321–332. https://doi.org/10.1007/s10745-017-9899-0

Gyau, A., Smoot, K., Diby, L., & Kouame, C. (2015). Drivers of tree presence and densities: The case of cocoa agroforestry systems in the Soubre region of Republic of Côte d'Ivoire. *Agroforestry Systems, 89*(1). https://doi.org/10.1007/s10457-014-9750-1

ICCO. (2022). *The International Cocoa Council adopts the Amended International Cocoa Agreement, 2010.* Retrieved on 7 May 2023, from https://www.icco.org/the-international-cocoa-council-adopts-the-amended-international-cocoa-agreement-2010/

Kalischek, N., Lang, N., Renier, C., Daudt, R. C., Addoah, T., Thompson, W., Blaser-Hart, W. J., Garrett, R., Schindler, K., & Wegner, J. D. (2022). Satellite-based high-resolution maps of cocoa planted area for Côte d'Ivoire and Ghana. arXiv:2206.06119

Knudsen, M. H., & Agergaard, J. (2015). Ghana's cocoa frontier in transition: The role of migration and livelihood diversification. *Geografiska Annaler: Series B, Human Geography, 97*(4), 325–342. https://doi.org/10.1111/geob.12084

Kolavalli, S., & Vigneri, M. (2011). Cocoa in Ghana: Shaping the Success of an Economy. In *Yes Africa can* (pp. 201–217). The World Bank. https://doi.org/10.1596/978-0-8213-8745-0

Kolavalli, S., & Vigneri, M. (2017). *The cocoa coast: The board-managed cocoa sector in Ghana.* International Food Policy Research Institute (IFPRI). https://doi.org/10.2499/9780896292680

Löwe, A. (2017). *Creating opportunities for young people in Ghana's cocoa sector.* Overseas Development Institute.

Meijer, S. S., Catacutan, D., Ajayi, O. C., Sileshi, G. W., & Nieuwenhuis, M. (2015). The role of knowledge, attitudes and perceptions in the uptake of agricultural and agroforestry innovations among smallholder farmers in sub-Saharan Africa. *International Journal of Agricultural Sustainability, 13*(1), 40–54. https://doi.org/10.1080/14735903.2014.912493

Ministry of Manpower Youth and Employment. (2007). *Labour practices in cocoa production in Ghana* (Issue April). Ministry of Manpower Youth and Employment.

Morris, E. K., Caruso, T., Buscot, F., Fischer, M., Hancock, C., Maier, T. S., Meiners, T., Müller, C., Obermaier, E., Prati, D., Socher, S. A., Sonnemann, I., Wäschke, N., Wubet, T., Wurst, S., & Rillig, M. C. (2014). Choosing and using diversity indices: Insights for ecological applications from the German Biodiversity Exploratories. *Ecology and Evolution, 4*(18), 3514–3524. https://doi.org/10.1002/ece3.1155

Mugure, A., Oino, P. G., & Sorre, B. M. (2013). Land ownership and its impact on adoption of agroforestry practices among rural households in Kenya: A case of Busia county. *International Journal of Innovation and Applied Studies, 4*(3), 552–559.

Nair, P. R., & Nair, V. D. (2014). 'Solid–fluid–gas': The state of knowledge on carbon-sequestration potential of agroforestry systems in Africa. *Current Opinion in Environmental Sustainability, 6*, 22–32.

Nunoo, I., Obiri, B. D., Frimpong, B. N., Isaac, N., Obiri, B. D., & Benedicta, N. F. (2015, September). *From deforestation to afforestation: Evidence from cocoa agroforestry systems.* XIV World Forestry Congress, pp. 7–11.

Nunoo, I., & Owusu, V. (2017). Comparative analysis on financial viability of cocoa agroforestry systems in Ghana. *Environment, Development and Sustainability, 19*(1), 83–98. https://doi.org/10.1007/s10668-015-9733-z

Obiri, B. D., Bright, G. A., McDonald, M. A., Anglaaere, L. C. N., & Cobbina, J. (2007). Financial analysis of shaded cocoa in Ghana. *Agroforestry Systems, 71*(2), 139–149. https://doi.org/10.1007/s10457-007-9058-5

Ongolo, S., Kouassi, S. K., Chérif, S., & Giessen, L. (2018). The tragedy of forestland sustainability in postcolonial Africa: Land development, Cocoa, and politics in Côte d'Ivoire. *Sustainability, 10*(12), 1–17. https://doi.org/10.3390/su10124611

Roth, M., Antwi, Y., O'Sullivan, R., & Sommerville, M. (2018). *Improving tenure security to support sustainable cocoa: Final report & lessons learned.* USAID Tenure and Global Climate Change Program.

Ruf, F., & Schroth, G. (2004). Chocolate forests and monocultures: A historical review of cocoa growing and its conflicting role in tropical deforestation and forest conservation. In G. Schroth, A.-M. N. Izac, H. L. Vasconcelos, C. Gascon, G. A. B. da Fonseca, & C. A. Harvey (Eds.), *Agroforestry and biodiversity conservation in tropical landscapes* (Vol. 06). Island Press. https://doi.org/10.1017/CBO9781107415324.004

Sadhu, S., Kysia, K., Onyango, L., Zinnes, C., Lord, S., Monnard, A., & Arellano, I. R. (2020, October). *NORC final report: Assessing progress in reducing child labor in cocoa production in cocoa growing areas of Côte d'Ivoire and Ghana* (Issue October). NORC at the University of Chigago. Retrieved on 20 August 2021, from https://www.norc.org/PDFs/Cocoa%20Report/NORC%202020%20Cocoa%20Report_English.pdf

Schroth, G., Krauss, U., Gasparotto, L., Aguilar, J. D., & Vohland, K. (2000). Pests and diseases in agroforestry systems of the humid tropics. *Agroforestry Systems, 50*(3), 199–241. https://doi.org/10.1023/A:1006468103914

Sellare, J., Meemken, E., & Qaim, M. (2020, February). Fairtrade, agrochemical input use, and effects on human health and the environment. *Ecological Economics, 176*, 106718. https://doi.org/10.1016/j.ecolecon.2020.106718

Smith Dumont, E., Gnahoua, G. M., Ohouo, L., Sinclair, F. L., & Vaast, P. (2014). Farmers in Côte d'Ivoire value integrating tree diversity in cocoa for the provision of ecosystem services. *Agroforestry Systems, 88*(6), 1047–1066. https://doi.org/10.1007/s10457-014-9679-4

Tscharntke, T., Clough, Y., Bhagwat, S. A., Buchori, D., Faust, H., Hertel, D., Hölscher, D., Juhrbandt, J., Kessler, M., Perfecto, I., Scherber, C., Schroth, G., Veldkamp, E., & Wanger, T. C. (2011). Multifunctional shade-tree management in tropical agroforestry landscapes: A review. *Journal of Applied Ecology, 48*(3), 619–629. https://doi.org/10.1111/j.1365-2664.2010.019 39.x

Van Wee, B., & Roeser, S. (2013). Ethical theories and the cost–benefit analysis-based ex ante evaluation of transport policies and plans. *Transport Reviews, 33*(6), 743–760. https://doi.org/10.1080/01441647.2013.854281

World Cocoa Foundation. (2022). *African Cocoa Initiative Phase II (ACI II): End of project report*. https://www.worldcocoafoundation.org/wp-content/uploads/2018/08/FINAL-WCF-ACI-II-Final-Report.pdf

Open Access This chapter is licensed under the terms of the Creative Commons Attribution 4.0 International License (http://creativecommons.org/licenses/by/4.0/), which permits use, sharing, adaptation, distribution and reproduction in any medium or format, as long as you give appropriate credit to the original author(s) and the source, provide a link to the Creative Commons license and indicate if changes were made.

The images or other third party material in this chapter are included in the chapter's Creative Commons license, unless indicated otherwise in a credit line to the material. If material is not included in the chapter's Creative Commons license and your intended use is not permitted by statutory regulation or exceeds the permitted use, you will need to obtain permission directly from the copyright holder.

CHAPTER 6

Can Agroforestry Provide a Future for Cocoa? Implications for Policy and Practice

Mette Fog Olwig◉, *Richard Asare*◉, *Philippe Vaast*◉, *and Aske Skovmand Bosselmann*◉

Abstract Climate change is threatening cocoa production in Ghana, the world's second largest cocoa exporter. Yet, as we have shown in this book, the impacts of climate change must be understood in the context of the multiple socioeconomic and biophysical pressures facing cocoa farmers, including the conversion of farms for other land uses, increasing hired labor costs as well as pests and diseases. This final chapter summarizes

M. F. Olwig (✉)
Department of Social Sciences and Business, Roskilde University, Roskilde, Denmark
e-mail: mettefo@ruc.dk

R. Asare
International Institute of Tropical Agriculture (IITA), Accra, Ghana
e-mail: r.asare@cgiar.org

P. Vaast
UMR Eco & Sols, Centre de Coopération Internationale en Recherche

© The Author(s) 2024
M. F. Olwig et al. (eds.), *Agroforestry as Climate Change Adaptation*, https://doi.org/10.1007/978-3-031-45635-0_6

the book's overall findings on cocoa agroforestry as climate change adaption and points to ways forward in terms of policy, practice and research. Our findings suggest that a nuanced view of farmers, agroecosystems and sites is necessary and emphasize the need to study shade tree species and species diversity, in addition to shade levels, to optimize the sustainability of cocoa farming. We further suggest that it may not be possible to sustainably grow cocoa in marginal regions of the cocoa belt, where yields are lower and where agroforestry may be unable to mitigate the negative impacts of the adverse climate. Finally, we point to the importance of considering rights and access to trees, land, extension services and resources, and call for more multidisciplinary research on differently situated farmers' opportunities and needs.

Keywords Climate change adaptation · Plant species diversity · Cocoa farmers · Sustainability · Institutional landscape · Multidisciplinary research

6.1   The Future of Cocoa Farming

Cocoa farming in Ghana, the second largest producer of cocoa in the world, is facing multiple pressures. These include biophysical pressures from climate change, pests and diseases, and socioeconomic pressures, such as the conversion of cocoa farms for other land uses, such as gold mines, and the lack of interest in cocoa farming among the young, leading to an aging of the cocoa-farming population. This book takes its point of

---

Agronomique Pour Le Développement (CIRAD), Université Montpellier, Montpellier, France

World Agroforestry Center, Nairobi, Kenya

P. Vaast
e-mail: philippe.vaast@cirad.fr

A. S. Bosselmann
Department of Food and Resource Economics, University of Copenhagen, Frederiksberg, Denmark
e-mail: ab@ifro.ku.dk

departure in the challenges posed by climate change, but climate change impacts must be understood in the context of these other factors because they together influence the overall sustainability of cocoa farming.

Based on the CLIMCOCOA research project, this book has specifically investigated cocoa agroforestry as a climate change adaptation strategy. Cocoa agroforestry entails planting cocoa trees together with non-cocoa trees, plants and crops. Agroforestry has been advocated in the literature as a way to counteract the negative impacts of climate change by providing shade and micro-climate buffering. It is also encouraged for its ability to mitigate climate change through the above- and below-ground carbon sequestration resulting from tree planting. It can furthermore enhance household food security, as well as improve farmers' livelihoods by diversifying their incomes (e.g., Graefe et al., 2017; Ruf & Schroth, 2004). While findings from previous research have generally been positive, there have also been conflicting findings and concerns regarding the impact of cocoa agroforestry on cocoa yields, pests and diseases, and the overall costs and benefits accrued by the cocoa farmers (e.g., Graefe et al., 2017; Nunoo & Owusu, 2017; Smith Dumont et al., 2014). This book has contributed to further nuancing and substantiating the possibilities and challenges of cocoa agroforestry in times of climate change by:

1. analysing the impacts of climate change on the socio-economic and biophysical bases of cocoa systems in Ghana
2. examining the complex of plant species involved in cocoa agroforestry and their environmental and societal attributes across a climate gradient
3. investigating the social and institutional contexts within which cocoa agroforestry practices are introduced.

Our overall findings indicate that cocoa agroforestry can be a successful way forward for cocoa farming. However, they also show that to succeed, place-specific socioeconomic and biophysical factors must be considered, and that the implementation of cocoa agroforestry must involve cross-sector collaboration between, for example, the state, chiefs, churches, NGOs, the cocoa industry and cocoa-farming communities. In this concluding chapter, we summarize the overall findings of the book and point to ways forward in terms of policy, practice and research on cocoa agroforestry that could ensure the sustainability and future of cocoa farming.

6.2   KEY OVERALL FINDINGS OF THE BOOK

The findings that have been presented in this book have nuanced our understanding of cocoa agroforestry as a means of climate change adaptation. They provide a better understanding of how climate changes are likely to influence cocoa farming and whether agroforestry can mitigate this impact. They point to the importance of looking at not just shade levels, but also shade tree species and species diversity when studying cocoa-agroforestry systems. They also emphasize the need to pay careful attention to the complex socioeconomic and institutional landscapes in which cocoa-agroforestry systems are introduced.

6.2.1   Climate Change, Cocoa and Agroforestry

Research on cocoa shows that cocoa cultivation is very vulnerable to climate change (Ameyaw et al., 2018; Schroth et al., 2016). West Africa, primarily Côte d'Ivoire and Ghana, where two thirds of the world's cocoa is farmed, will experience an increasing frequency and severity of drought and heat (see Chapters 1 and 2). A key contribution of this book is its examination of how cocoa yields have responded historically to changes in climate and how this knowledge can help understand the impact of different future climate change scenarios on the sustainability of cocoa farming. Additionally, the book has investigated what exactly happens to the cocoa plant, i.e., to the plant's physiology, when it is exposed to different climate stressors, such as high temperatures and drought. Furthermore, the book has shown how the challenges and potentials of cocoa agroforestry vary under different climates in relation to both biophysical and socioeconomic outcomes.

We analyzed the relationship between historical cocoa yields and climate in Ghana across the six decades spanning 1960–2020 (see Chapter 1). Overall, the analysis showed that the levels and timing of both temperature and precipitation impacted yields. Annual cocoa production was positively correlated with precipitation in and around the major dry season, particularly in the month of November. Negative correlations were observed between cocoa production and temperatures in the minor dry season around July–August. The minor wet season from September to November coincides with the period when cocoa trees have many maturing pods. A limited water supply during this period can reduce photosynthesis and have a negative effect on pod yields (Asante et al.,

2022). Generally, there was a positive correlation between precipitation in the minor wet season and yields across major parts of the cocoa belt in Ghana, including the Western, Eastern, Central, Brong Ahafo, and Ashanti regions. In the Volta region, correlations were weak because production was low after the 1970s. This was in part due to devastating bushfires in the early 1980s, combined with the severe incidence of Cocoa Swollen Shoot Virus (CSSV), which almost wiped out the cocoa farms in the Volta region (Danquah, 2003).

Based on two trials, we documented how cocoa-plant physiology is influenced by increasing temperatures and reduced precipitation, and the mitigating effect of shade (see Chapter 2; also, Mensah et al., 2022). We first demonstrated how cocoa seedlings exposed to temperatures 5–7 °C above their surroundings had an increased risk of damage and reduced photosynthesis. Cacao plants under shade had thin leaves, which is a typical shade-leaf anatomy, and increased rates of photosynthesis. Under heat stress, shade was partially able to mitigate the damaging effects of high temperatures. In another experiment, we showed that mature cocoa trees exposed to reductions in rainfall were increasingly vulnerable to flower abortion and had substantially reduced yields. At all levels of rainfall reduction, cocoa shaded by a 40% shade net performed better in terms of yield compared to unshaded cocoa. This suggests that shade has a positive impact irrespective of water supply under these circumstances.

Comparing yields in cocoa-agroforestry systems along a climate gradient, our research shows that yields from cocoa-agroforestry systems decreased from the wet southern to the dry northern part of the cocoa belt of Ghana (Asitoakor et al., 2022b; Chapter 5). For our data collection and analysis,[1] building on Bunn et al. (2019), we divided the cocoa belt of Ghana into three climate impact zones: the Cope Zone, the Adjust Zone and the Transform Zone. The southern Cope Zone has a current climate that is the most favorable to cocoa farming of the three, and cocoa farming is likely to be able to *cope* with climate change. In the middle Adjust Zone, the current climate is moderately favorable, but some *adjustments* to cocoa farming will likely be needed. The northern Transform Zone currently has the climate that is least favorable to cocoa farming. Here, cocoa farms will likely have to be abandoned or radically *transformed* because of climate changes. Through a cost–benefit analysis

---

[1] See Chapter 1, as well as individual chapters, for a more in-depth discussion of our methods, as well as a map of the study sites.

of cocoa agroforestry based on household surveys in the three different climate impact zones (see Chapter 5), we found that the costs and benefits differed across the different climate impact zones. Cocoa bean yield per hectare was for example significantly higher in the Cope zone. In a different study, Abdulai et al. (2018b) found that in marginal regions of the cocoa belt, such as in the Transform Zone, cocoa agroforestry has a limited positive effect, and can even have a negative effect under conditions of severe drought. It is important to consider region-specific climate conditions and projections when selecting the appropriate level of shade and shade tree characteristics to implement a sustainable strategy to buffer climate change. In fact, several of our findings indicate that, as the climate continues to change, marginal areas become even less suitable for cocoa farming. It may not be cost-effective to continue cocoa production in these areas, especially since cocoa agroforestry does not appear to buffer these changes sufficiently and may in some cases even worsen climate impacts in such areas.

6.2.2   *The Importance of Shade Tree Species and Species Diversity*

Another contribution of this book is its focus on the different effects of shade tree species and tree species diversity. Within agroforestry research on cocoa and coffee, there has been a tendency to focus narrowly on the impact of different levels of shade, with little regard to which constellation of trees is providing this shade. However, new research emphasizes that the effects on cocoa may depend on the particular shade tree species involved, as well as the impact of shade tree species diversity (Abdulai et al., 2018a; Asare et al., 2019; Asitoakor et al., 2022a; Graefe et al., 2017; Kaba et al., 2020). Thus, there is a need for broader investigations of (1) the impact of specific shade tree species, and (2) agroecosystem studies that include a focus on viruses, fungi, animals and plants. In coco-agroforestry systems, the foliage density, the root distribution along the soil profile and the associated below-ground complementarity and competition for resources are of key importance and will be influenced differently by different shade tree species (Abdulai et al., 2018b; Critchley et al., 2022; Jaimes-Suarez et al., 2022). Important factors include the depth of the shade tree's root system (Kyereh, 2017) and the water requirements of the species involved across space and time, influencing competition between shade trees and cocoa (Abdulai et al., 2018b; Adams et al., 2016).

Our trial experiments showed that climate stressors negatively impact cocoa-plant physiology, and that shade has a positive impact both under stress and no-stress conditions (see Chapter 2; also Mensah et al., 2022). A limitation of these trial experiments is that they used shade nets, not shade trees, to achieve a uniform shade cover. To investigate the significance of choosing different shade tree species, we set up a farm study experiment comparing how eight common forest shade trees species affected cocoa trees and their yields, as well as the impact of mirid insects and black pod disease (see Chapter 3). Although this on-farm study experiment is just a first step in this field of study, the findings indicate that some shade tree species significantly outperformed the full-sun control plot with respect to yields and the occurrence of pests and diseases. Previous thinking has been that shade trees reduce yields and enhance the incidence of pests and diseases, particularly under high input conditions and with high-quality cocoa-planting material. However, a recent literature review (Mattalia et al., 2022) challenges this assumption, and our findings furthermore suggest that yields can be higher from shaded cocoa compared to full-sun cocoa, especially under a low input of fertilizer, insecticides and fungicides, if the right shade tree species are selected for the local context.

The impact of the level of tree species diversity (and not just the level of shade) on the various costs and benefits associated with cocoa agroforestry was also explored in our research (see Chapter 5). We found that cocoa agroforestry was more profitable than monocrop systems when combined with income from the sale of other products from diverse agroforestry systems, such as timber, fuelwood, fruit and mushrooms. Moreover, cocoa farmers earned a consistent income from their cocoa plots if they included more tree species in their system. The need for hired labor (e.g., related to manual weeding and applying inputs) was higher for cocoa plots with low tree species diversity compared to those with medium tree species diversity. This finding is important because one of the key concerns regarding the future of cocoa is that it is highly labor-intensive and therefore unattractive to young people with other aspirations and alternative livelihood possibilities (Anyidoho et al., 2012). In addition to weeding and the application of inputs, cocoa farming involves pruning, harvesting, gathering and breaking pods, fermenting, transporting and drying cocoa beans. This can lead to illegal solutions, such as using child labor, to reduce the expenses incurred from hiring adult laborers. The problems of youth disinterest and the use of illegal child labor could potentially be addressed if diverse cocoa-agroforestry systems that are less labor-intensive were to be adopted.

6.2.3   Access and Rights

Research may provide recommendations concerning shade levels and the need for fertilizers and insecticides, as well as the planting, timely harvesting and pruning of cocoa-agroforestry systems. However, none of these recommendations can be implemented in practice if farmers do not have access to seedlings, extension services and key inputs, along with long-term rights to the land and trees (Boadi et al., 2022). A final key overall contribution of this book is to illuminate the importance of the socioeconomic and institutional factors that directly and indirectly influence the outcomes of cocoa-agroforestry systems in relation to benefits to both farmers' livelihoods and the environment.

Farmers' own perspectives were explored through twenty focus-group discussions and interviews (see Chapter 4). Farmers were found to generally agree on the possible benefits of having shade trees in cocoa cultivation. The benefits ranged from creating a better environment for the cocoa trees at different stages of the cocoa plot's lifetime to being able to harvest alternative products, including snails and mushrooms living in the shaded environment, and products from the trees (see Chapter 3, Appendix for a list of common shade tree species adopted in cocoa-agroforestry systems and their additional uses). Despite this common knowledge, most farmers establish new plantations by clearcutting and burning fallow or forested areas, and while some introduce new shade trees, a widespread adoption of agroforestry systems is lacking, as farmers experience a range of obstacles and challenges. One of the main challenges is farmers' access and rights to land, as well as to the trees on the land. Village chiefs have the constitutional right and duty to administer land in the interest of the community (1992 Constitution, article 36(8)). However, many farmers, who were sharecroppers and whose families had migrated to the village several generations previously, complained that they are often still perceived as outsiders, and found that the chiefs used their positions to expropriate their cocoa fields and replace them with, for example, urban expansion, village infrastructure or sand mining. Other farmers who had good relations with their chief, or who possessed ancestral rights of ownership to their land, did not worry about their future ability to access their land. They were therefore more willing to invest in cocoa agroforestry, even though the benefits of planting trees will only accrue after a number of years. Access and rights to land vary between and within regions, villages and even households, with women being a lot less likely to own cocoa land than men (Barrientos & Bobie, 2016).

However, regardless of land rights, farmers do not have the right to fell timber trees they have planted or nurtured on their farm unless they can prove ownership and secure a permit from the Forestry Commission of Ghana. As permits are difficult to obtain because of the bureaucracy involved, several farmers had experienced legal conflicts with forestry personnel over the use of trees, even for their own housing materials, and they therefore saw few incentives to continue caring for trees. Another significant competitor in certain areas is gold mining, which leaves land unusable for cocoa farming, and provides a lucrative alternative livelihood for young people which contributes to their loss of interest in cocoa farming. Several actors come into play here, such as the mining companies that encourage small-scale mining activities in the cocoa communities. Farmers, mostly representing the older generation, talked of defending their lands against outside gold miners and discouraging their own children from engaging in mining.

To successfully implement agroforestry systems, cocoa farmers must have knowledge of the different appropriate shade tree species, as well as the other plants involved. We found that the more resources and better networks farmers had, and thus the easier access to inputs and knowledge of good management practices, the more likely they were to implement agroforestry systems (see Chapter 5). Another significant finding pertains to the role played by the extension services provided not just by government agencies, but also external institutions like NGOs, research organizations and businesses, such as those tied to the cocoa and chocolate industry. While cocoa farmers learn from each other as they see how other farmers manage their cocoa farms, our study found that those farmers who were most successful in implementing cocoa agroforestry had received assistance from extension services (see Chapter 4). This assistance included advice as well as concrete inputs, such as a more diverse selection of shade tree seedlings. However, most farmers receive only limited training, and only a very limited selection of shade tree species is available from most NGOs, cocoa industry-led sustainability initiatives, or the state. This is especially a problem for women, because they are often not recognized officially as cocoa farmers. Studies show that 80% of registered cocoa farmers are men, even though women carry out close to half of the cocoa work required on farms as unpaid family labor. Because women are unregistered as cocoa farmers, they are often not included in training and do not receive extension services (Barrientos & Bobie, 2016).

Local botanical knowledge is instrumental in the adoption of cocoa-agroforestry practices because farmers can diversify the shade tree species on their cocoa farms through naturally regenerated trees or tree seedlings acquired from other farmers (Rigal et al., 2022). Furthermore, to be successful, both economically and ecologically, cocoa agroforestry necessitates different constellations of non-cocoa tree species and other plants at different times, depending on the height and age of the cocoa trees. Cocoa-agroforestry systems are in effect three-dimensional arrangements of trees and plants: on the ground, in the canopy and under the soil (Asare, 2006) with time constituting a fourth dimension.

6.3   Implications for Policy and Practice

To ensure a more sustainable production of cocoa, several global and national initiatives and policies have been put in place. The Cocoa & Forests Initiative was introduced by the cocoa and chocolate sector in 2017 as a collective commitment to address deforestation and forest degradation in the cocoa supply chain, focusing initially on Ghana and Côte d'Ivoire. Yet, questions have been raised regarding the effectiveness of such voluntary sustainability measures, and calls have been made for the coordinated accountability of public and private activities (Caro-denuto & Buluran, 2021). Another initiative is the Emission Reductions Payment Agreements (ERPAs) for the Carbon Fund with the World Bank as a Trustee, signed by the government of Ghana in 2019 (ER-MR, 2021). Under this program, there is a benefit-sharing plan that guides the sharing of Carbon Benefits generated under the Ghana Cocoa Forest REDD+ Program (GCFRP). The GCFRP uses a climate-smart cocoa-production strategy, which is the world's first commodity-based emission reductions program that aims to significantly reduce deforestation and forest degradation-driven emissions, while making sure that smallholders' livelihoods are improved through increases in yields.

A further initiative is the Living Income Differential (LID) policy from 2019, which the cocoa marketing boards of Ghana and Côte d'Ivoire have established with the chocolate companies. The Ghanaian cocoa marketing board (The Ghana Cocoa Board, or Cocobod) regulates the pricing, purchasing, marketing and exportation of cocoa beans in Ghana, and provides support programs to farmers (see Chapter 5). The aim of the LID policy is to add a premium to the price of cocoa to ensure a living income for cocoa farmers, defined as the "net annual income required for a household in a particular place to afford a decent standard of living for all members of that household" (Adams & Carodenuto, 2023, p. 2). It is envisaged that GCFRP and LID together will make Ghana's cocoa and forestry sectors more resilient with earnings from climate-smart cocoa beans that promote the active incorporation of shade trees when establishing new or rehabilitating old plantations. However, studies have pointed to the need for such policies to give greater consideration to farmer diversity in relation to, for example, tenure, farm size and management strategies (Adams & Carodenuto, 2023).

While implementing policies that address the broader institutional landscape, along with land use, is of key importance, this is difficult and takes time. Therefore, in the short term, the institutional challenges farmers face must be taken into consideration by policymakers, practitioners and researchers when researching and implementing cocoa-agroforestry systems. In the following, we provide recommendations for policy and practice in relation to how to optimize the complex of plant species involved in a cocoa-agroforestry system while being mindful of the socioeconomic and institutional landscape. These recommendations can broadly be organized into three categories depending on whether they relate to: (1) the components going into the system, (2) how the system functions, and (3) the outputs of the system.

6.3.1   *The Components Going into the Cocoa-Agroforestry System*

When cocoa was first introduced to Ghana in the 1880s, it was established as an unplanned agroforestry system that depended on forest-fallow regimes and their natural processes of regeneration, thus enabling the farmers to organize and diversify cocoa-agroforestry systems (Asare & Asare, 2008). Shade trees were maintained either because they were deemed important or because farmers did not have the equipment needed to fell them, and the land was then planted with cocoa seedlings, food and cash crops to provide shade for the seedlings and to obtain food and income (Osei-Bonsu et al., 1998). More recently, cocoa farms have been established by completely clearing the land through felling and burning, after which farmers plant shade trees and food crops followed by cocoa seedlings. Farmers also remove regenerated forest-tree saplings that are seen as competing with the cocoa seedlings while nurturing those that are believed to be of value (Asare & Asare, 2008).

As we have shown, to implement a cocoa-agroforestry system, seedlings and other plants must be obtained, labor is needed, and various inputs such as fertilizer and insecticide must be applied. Time is also a factor, specifically how much time the farmer can put into the system before outcomes are required. This will in part depend on farmers' rights and access to the land. Extension services that provide guidance on how and when to plant, prune, harvest and apply inputs can affect outcomes. What constitutes an optimal complex of plant species depends on these different elements. Thus, if a farmer has little time and labor, and no access to inputs or regular extension services, a different complex of plant species will be more advantageous than is the case for a farmer who has a longer time horizon, can afford paid labor or has family labor available, and who has easy access to inputs and services. The ability to access these different elements varies from place to place and from farmer to farmer, hence it is important for policymakers and practitioners to understand the local context when seeking to support and implement cocoa agroforestry. That said, in general, many farmers in Ghana operate with short time horizons, need to minimize labor and other inputs, and do not have easy access to seedlings or extension services (Boadi et al., 2022). Furthermore, current tree-tenure arrangements specific to the Ghanaian regulatory context require farmers to register the trees on their farms with the Forestry Commission. This effectively creates an unnecessary disincentive for farmers to care for trees. We therefore recommend that research, policy and practice focus on how to optimize the complex of plant species for this group of farmers—biophysically, socio-ecologically and in terms of regulations. Overall, our findings that are relevant in this regard indicate that shade reduces the need for inputs and that greater shade tree species diversity in cocoa-agroforestry systems reduces labor needs.

6.3.2   How the Cocoa-Agroforestry System Functions

In addition to being aware of the different components going into the cocoa-agroforestry system, it is important to pay attention to how the complex of plant species influences system processes. These processes include competition and complementarity between species (both in terms of water and nutrients) and the occurrence of pests and diseases. Moreover, shade levels will affect the climate resilience of the system (Asitoakor et al., 2022a, 2022b; Mensah et al., 2022). These processes vary between climate zones and sites, and we therefore recommend that this must be considered by researchers, policymakers and practitioners. Our findings that are relevant in this regard indicate that implementing cocoa agroforestry sustainably in the Transform Zone, the zone with the least suitable climate for cocoa farming, may not be possible. Agroforestry was unable to mitigate the negative impacts of the adverse climate and could in fact have a negative effect. Besides, as shown in our cost–benefit analysis, cocoa agroforestry in this zone produced the lowest yields.

It should be pointed out, however, that with more research it may be possible to identify shade tree species that better buffer the negative impacts of climate change in climates corresponding to the Transform Zone. This, coupled with breeding for drought-resistant cocoa, could make cocoa farming in marginal areas viable. Nevertheless, our results indicate that resources would be better utilized if research efforts were focused on sustainable cocoa farming in the other two zones, in particular the Cope Zone. In these two zones, our findings show, shade leads to a more optimal plant physiology under stress conditions instigated by changes in the climate. Impacts depend, however, on complementarity in water use between shade tree species and cocoa. Deep-rooted shade trees that tap soil water below the cocoa root zone may work best. Furthermore, different shade tree species appear to lead to different levels of pest and disease incidence.

6.3.3 Outputs from the Cocoa-Agroforestry System

The output of the cocoa-agroforestry system can be assessed in terms of improvements to the productivity of cocoa beans, fruit and timber as well as the lifespan of the cocoa trees, and in terms of the possibilities for on-farm or off-farm diversification. Research has shown that shade can prolong the economic lifespan of cocoa trees (Obiri et al., 2007), and our findings indicate that different shade tree species may affect overall yields differently. For the farmer, the degree to which the benefits outweigh the costs of the cocoa-agroforestry system does not just depend on the yield of cocoa. The integration of trees and crops that provide marketable products such as timber and fruit is also important.

If done correctly, cocoa agroforestry can be more consistently profitable, while being less labor-intensive than monocrop systems. This could free up labor for other livelihood activities, reduce the use of illegal child labor and make cocoa farming a more attractive option for the young. We therefore recommend that farmers are supported in diversifying cocoa-agroforestry systems in terms of shade tree species and involving fruits and other products for household consumption and sale. This includes making the best use of easily available self-sown shade tree seedlings to maximize outputs of foodstuffs and timber for household consumption and sale. It is also important to encourage farmers to choose shade tree species that increase cocoa yields while reducing pests and diseases without competing with cocoa trees for water and nutrients. Importantly, extension services should pay particular attention to cocoa farmers that are not officially registered and therefore easily bypassed, often women and migrants that are less likely to own cocoa land yet carry out a significant proportion of cocoa work.

6.4   Moving Forward

The future of cocoa is unclear. As a result of climate change, diseases and weather variations, supply deficits are projected (ICCO, 2023). The impact of supply deficits on cocoa management, prices and quality requirements is uncertain. One possible response is an intensification of cocoa farming that involves low to no shade trees and a high need for agricultural inputs. However, this will come at the expense of the environment and likely the small-scale farmers. New regulations on the world's main chocolate market—the EU—have set new requirements for companies that import and trade cocoa beans and their derivatives. From 2025, importers must document that the products are not associated with deforestation or forest degradation, among other environmental concerns. While this may not affect areas deforested before 2021, it is expected to influence the expansion of new cocoa areas (Li et al., 2022). Moreover, it may increase the number of sustainability projects licensed cocoa-buying companies conduct in producing countries. It is anticipated that these projects will focus on the dissemination of shade trees, as well as information on environmental and social issues. Demand for quality cocoa is also expanding rapidly, with Ghana and Côte d'Ivoire poorly positioned because cocoa beans originating from these countries are of far lower organoleptic (sensory) quality than beans from Latin America (Fountain & Hütz-Adams, 2022). There is an increasing interest in assessing the effect of shade trees on cocoa quality. This book argues that agroforestry can address many of the challenges currently faced by cocoa farmers in Ghana, and more broadly in West Africa, and therefore provides a sustainable future pathway for cocoa. Nevertheless, more research is needed to better understand and implement cocoa agroforestry.

6.4.1   More Focus on Shade Tree Species

In contemporary research, policy and practice, there is a global tendency to regard and present tree planting as a straightforward and inexpensive panacea ameliorating climate change. The findings presented in this book offer an informed alternative to this simplistic approach to, and understanding of, tree planting. We have stressed that to increase cocoa farmers' engagement in cocoa agroforestry and make cocoa farming more sustainable, place-specific knowledge concerning the effects of shade tree species is needed. It is time to move beyond the generic focus on shade levels and cocoa yields.

We have suggested that research needs to place particular emphasis on how to minimize the need for inputs, including time, labor and fertilizers. One potential avenue for research in this regard is to look at the relationship between specific shade tree species and the need for inputs in cocoa farming. Some shade tree species will, for example, host pests that would otherwise have concentrated on the cocoa tree. Research is needed to understand the net effect of shade trees on the incidence of pests on cocoa trees, which impacts the need for chemicals. Another important avenue for research is to investigate the impacts of the age of both shade trees and cocoa trees.

6.4.2   Multidisciplinary Research

The book has investigated the potential of cocoa agroforestry in times of climate change, focusing not just on cocoa yields, but also on how to ensure that cocoa farming remains a viable and attractive livelihood option for farmers, including future generations of farmers. This has only been possible by employing multidisciplinary approaches. Unfortunately, in practice, such approaches are difficult to implement for several reasons. Researchers may not be used to communicating across disciplines, and it is difficult to receive funding for multidisciplinary approaches, which can involve greater expense because different research methods can require different equipment and research set-ups. In relation to research on trees and climate change, studies need to be longitudinal to obtain results, and many funding bodies will only fund a maximum of three to five years of research.

In the CLIMCOCOA research project, we worked deliberately on ensuring communication across disciplines. This was done in different ways, first by associating researchers from different disciplines from the beginning of the study design. We also organized reading groups where we discussed texts from different disciplines and presented our research and approaches to other team members, as well as to a broader audience, for example through conference panels where both the biophysical and socioeconomic findings were presented. This enabled us to identify findings that cut across both the biophysical and the socioeconomic, such as our finding that it may not be possible to implement cocoa agroforestry sustainably in marginal regions of the cocoa belt (e.g., the Transform zone).

Perhaps the most important outcome of multidisciplinary research is that it enables researchers to understand and communicate research topics in a larger context, rather than focusing on a narrow research agenda. Thereby, the societal relevance of the research becomes greater. In this sense, the present book represents a step toward a better understanding of the interrelations of biophysical and socioeconomic factors and points to the need for further multidisciplinary research on climate change, sustainability and agriculture.

6.4.3   *More Focus on Farmers*

In recent years, research on the sustainability of cocoa farming has expanded significantly. This is not only due to the threats caused by climate change, but also because cocoa and chocolate consumers and investors are increasingly expecting the cocoa industry to address sustainability concerns in the cocoa sector. This includes both environmental and social aspects, such as acceptable working conditions and the elimination of child labor. This has created an opportunity for researchers, civil society, NGOs and policymakers to direct research activities toward sustainable cocoa farming and challenged the chocolate industry to be actively involved in research on sustainability in the cocoa sector.

We welcome this concern for the sustainability of farming cocoa, which has been studied less than other cash crops such as coffee. However, it is important that research is not only site- and species-specific, but that it also considers that farmers are individuals with different options and interests. This includes paying particular attention to farmers who are systematically underrepresented in research on cocoa farming because they are not the official landowners, or are not officially registered as cocoa farmers, such as women and migrants. At present, and given the growing age of cocoa farmers, the lack of interest among the young may be one of the biggest threat to the future of cocoa in Ghana. It is therefore of crucial importance that research on cocoa agroforestry not only examines the climate resilience of the agroecosystem, but also systematically investigates the vital role of both socioeconomic and institutional factors and concerns. Without farmers, there will be no cocoa farming.

References

Abdulai, I., Jassogne, L., Graefe, S., Asare, R., Van Asten, P., Läderach, P., & Vaast, P. (2018a). Characterization of cocoa production, income diversification and shade tree management along a climate gradient in Ghana. *PLoS ONE, 13*(4), 1–17. https://doi.org/10.1371/journal.pone.0195777

Abdulai, I., Vaast, P., Hoffman, M., Asare, R., Jassogne, L., Asten, V. P., Rotter, P. R., & Graefe, S. (2018b). Cocoa agroforestry is less resilient to sub-optimal and extreme climate than cocoa in full sun. *Global Change Biology, 24*(1), 273–286.

Adams, M. A., Turnbulla, T. L., Sprent, J. I., & Buchmannc, N. (2016). Legumes are different: Leaf nitrogen, photosynthesis, and water use efficiency. *PNAS, 113*, 4098–4113.

Adams, M. A., & Carodenuto, S. (2023). Stakeholder perspectives on cocoa's living income differential and sustainability trade-offs in Ghana. *World Development, 165.* https://doi.org/10.1016/j.worlddev.2023.106201

Ameyaw, L. K., Ettl, G. J., Leissle, K., & Anim-Kwapong, G. J. (2018). Cocoa and climate change: Insights from smallholder cocoa producers in Ghana regarding challenges in implementing climate change mitigation strategies. *Forests, 9*(12), 742.

Anyidoho, N. A., Leavy, J., & Asenso-Okyere, K. (2012). Perceptions and aspirations: A case study of young people in Ghana's cocoa sector. *IDS Bulletin, 43*(6), 20–32. https://doi.org/10.1111/j.1759-5436.2012.00376.x

Asante, P. A., Rahn, E., Zuidema, P. A., Rozendall, D. M. A., van der Baan, M. E. G., Läderach, P., Asare, R., Cryer, N. C., & Anten, N. P. R. (2022). The cocoa yield gap in Ghana: A quantification and an analysis of factors that could narrow the gap. *Agricultural Systems, 201*, 103473. https://doi.org/10.1016/j.agsy.2022.103473

Asare, R. (2006). *Learning about neighbour trees in cocoa growing systems—A manual for farmer trainers* (Development and Environment Series 4-2006). Forest & Landscape Denmark.

Asare, R., & Asare, A. R. (2008, September). *A participatory approach for tree diversification in cocoa farms: Ghanaian farmers' experience* (STCP Working Paper Series 9). International Institute of Tropical Agriculture.

Asare, R., Markussen, B., Asare, R. A., Anim-Kwapong, G., & Ræbild, A. (2019). On-farm cocoa yields increase with canopy cover of shade trees in two agro-ecological zones in Ghana. *Climate and Development, 11*(5), 1–12. https://doi.org/10.1080/17565529.2018.1442805

Asitoakor, B. K., Asare, R., Ræbild, A., Ravn, H. P., Eziah, V. Y., Owusu, K., Mensah, E. O., & Vaast, P. (2022b). Influences of climate variability on cocoa health and productivity in agroforestry systems in Ghana. *Agricultural and Forest Meteorology, 327*(109199), 1–13. https://doi.org/10.1016/j.agrformet.2022.109199

Asitoakor, B. K., Vaast, P., Ræbild, A., Ravn, H. P., Eziah, V. Y., Owusu, K., Mensah, E. O., & Asare, R. (2022a). Selected shade tree species improved cocoa yields in low-input agroforestry systems in Ghana. *Agricultural Systems, 202*(103476), 1–9. https://doi.org/10.1016/j.agsy.2022.103476

Barrientos, S., & Bobie, A. O. (2016). *Promoting gender equality in the cocoa-chocolate value chain: Opportunities and challenges in Ghana* (GDI Working Paper 2016-006). University of Manchester.

Boadi, S. A., Olwig, M. F., Asare, R., Bosselmann, A. S., & Owusu, K. (2022). The role of innovation in sustainable cocoa cultivation: Moving beyond mitigation and adaptation. In M. Coromaldi & S. Auci (Eds.), *Climate-induced innovation: Mitigation and adaptation to climate change* (pp. 47–80). Springer.

Bunn, C., Fernandez-Kolb, P., Asare, R., & Lundy, M. (2019, September). *Climate smart cocoa in Ghana: Towards climate resilient production at scale.* CCAFS Info Note. https://cgspace.cgiar.org/bitstream/handle/10568/103770/CCAFS%20Info%20Note_%20cocoa%20Ghana%20finalized_20190930.pdf

Carodenuto, S., & Buluran, M. (2021). The effect of supply chain position on zero-deforestation commitments: Evidence from the cocoa industry. *Journal of Environmental Policy & Planning, 23*(6), 716–731. https://doi.org/10.1080/1523908X.2021.1910020

Critchley, M., Sasse, M., Rahn, E., Ashiagbior, G., Soesbergen, A., & Maney, C. (2022). *Identifying opportunity areas for cocoa agroforestry in Ghana to meet policy objectives.* United Nations Environmental Programme of World Conservation Monitoring Centre.

Danquah, F. K. (2003). Sustaining a West African cocoa economy: Agricultural science and the Swollen Shoot contagion in Ghana, 1936–1965. *African Economic History* (31), 43–74.

ER-MR. (2021). Forest Carbon Partnership Facility (FCPF) carbon fund ER Monitoring Report (ER-MR), Forestry Commission.

Fountain, A. C., & Hütz-Adams, F. (2022). *2022 cocoa barometer.*

Graefe, S., Meyer-Sand, L. F., Chauvette, K., Abdulai, I., Jassogne, L., Vaast, P., & Asare, R. (2017). Evaluating farmers' knowledge of shade trees in different cocoa agro-ecological zones in Ghana. *Human Ecology, 45*(3), 321–332. https://doi.org/10.1007/s10745-017-9899-0

ICCO. (2023.) *Quarterly Bulletin of Cocoa Statistics*, Vol. XLIX, No.1, Cocoa year 2022/2023.

Jaimes-Suarez, Y. Y., Carvajal-Rivera, A. S., Galvis-Neira, D. A., Carvalho, F. E. L., & Rojas-Molina, J. (2022). Cacao agroforestry systems beyond the stigmas: Biotic and abiotic stress incidence impact. *Frontiers in Plant Science, 13*, 921469.

Kaba, J. S., Otu-nyanteh, A., & Abunyewa, A. A. (2020). The role of shade trees in influencing farmers' adoption of cocoa agroforestry systems: Insight from semi-deciduous rain forest agroecological zone of Ghana. *NJAS—Wageningen Journal of Life Sciences, 92*(100332), 1–7. https://doi.org/10.1016/j.njas.2020.100332

Kyereh, D. (2017). Shade trees in cocoa agroforestry systems in Ghana: Influence on water and light availability in dry seasons. *Journal of Agriculture and Ecology Research International, 10*(2), 1–7.

Li, B., Schneider, T., Stolle, F., &. Veldhoven, S. (2022). *How a new EU regulation can reduce deforestation globally.* World Resources Institute. https://www.wri.org/insights/eu-deforestation-regulation

Mattalia, G., Wezel, A., Costet, P., Jagoret, P., Deheuvels, O., Migliorini, P., & David, C. (2022). Contribution of cacao agroforestry versus mono-cropping systems for enhanced sustainability. A review with a focus on yield. *Agroforestry Systems, 96*(1). https://doi.org/10.1007/s10457-022-00765-4

Mensah, E. O., Asare, R., Vaast, P., Amoatey, C. A., Markussen, B., Owusu, K., Asitoakor, B. K., & Ræbild, A. (2022). Limited effects of shade on physiological performances of cocoa (Theobroma cacao L.) under elevated temperature. *Environmental and Experimental Botany, 201*(104983), 1–11.

Nunoo, I., & Owusu, V. (2017). Comparative analysis on financial viability of cocoa agroforestry systems in Ghana. *Environment, Development and Sustainability, 19*(1), 83–98. https://doi.org/10.1007/s10668-015-9733-z

Obiri, B. D., Bright, G. A., McDonald, M. A., Anglaaere, L. C., & Cobbina, J. (2007). Financial analysis of shaded cocoa in Ghana. *Agroforestry Systems, 71*, 139–149.

Osei-Bonsu, K., Amoah, F. M., & Oppong, F. K. (1998). The establishment and early yield of cocoa intercropped with food crops in Ghana. *Ghana Journal of Agricultural Science, 31,* 45–53.

Rigal, C., Wagner, S., Nguyen, M. P., Jassogne, L., & Vaast, P. (2022). ShadeTreeAdvice methodology: Guiding tree-species selection using local knowledge. *People and Nature,* 16 pp. https://doi.org/10.1002/pan3.10374

Ruf, F., & Schroth, G. (2004). Chocolate forests and monocultures: A historical review of cocoa growing and its conflicting role in tropical deforestation and forest conservation. In G. Schroth, A.-M. N. Izac, H. L. Vasconcelos, C. Gascon, G. A. B. da Fonseca, & C. A. Harvey (Eds.), *Agroforestry and biodiversity conservation in tropical landscapes* (Vol. 06). Island Press. https://doi.org/10.1017/CBO9781107415324.004

Schroth, G., Läderach, P., Martinez-Valle, A. I., Bunn, C., & Jassogne, L. (2016). Vulnerability to climate change of cocoa in West Africa: Patterns, opportunities, and limits to adaptation. *Science of the Total Environment, 556,* 231–241.

Smith Dumont, E., Gnahoua, G. M., Ohouo, L., Sinclair, F. L., & Vaast, P. (2014). Farmers in Côte d'Ivoire value integrating tree diversity in cocoa for the provision of ecosystem services. *Agroforestry Systems, 88*(6), 1047–1066. https://doi.org/10.1007/s10457-014-9679-4

Open Access This chapter is licensed under the terms of the Creative Commons Attribution 4.0 International License (http://creativecommons.org/licenses/by/4.0/), which permits use, sharing, adaptation, distribution and reproduction in any medium or format, as long as you give appropriate credit to the original author(s) and the source, provide a link to the Creative Commons license and indicate if changes were made.

The images or other third party material in this chapter are included in the chapter's Creative Commons license, unless indicated otherwise in a credit line to the material. If material is not included in the chapter's Creative Commons license and your intended use is not permitted by statutory regulation or exceeds the permitted use, you will need to obtain permission directly from the copyright holder.

INDEX

© The Editor(s) (if applicable) and The Author(s) 2024
M. F. Olwig et al. (eds.), *Agroforestry as Climate Change Adaptation*,
https://doi.org/10.1007/978-3-031-45635-0